Contents

Figures

Tables

Appendixes

Contacts

Summary

Nanotechnology—a term encompassing the science, engineering, and applications of submicron materials—involves the harnessing of unique physical, chemical, and biological properties of nanoscale substances in fundamentally new and useful ways. The economic and societal promise of nanotechnology has led to substantial and sustained investments by governments and companies around the world. In 2000, the United States launched the world's first national nanotechnology program. From FY2001 through FY2012, the federal government invested approximately $15.6 billion in nanoscale science, engineering, and technology through the U.S. National Nanotechnology Initiative (NNI). President Obama has requested $1.8 billion in NNI funding for FY2013. U.S. companies and state governments have invested billions more. As a result of this focus and these investments, the United States has, in the view of many experts, emerged as a global leader in nanotechnology. However, the competition for global leadership in nanotechnology is intensifying as countries and companies around the world increase their investments.

Nanotechnology's complexity and intricacies, early stage of development (with commercial pay-off possibly years away for many potential applications), and broad scope of potential applications engender a wide range of public policy issues. Maintaining U.S. technological and commercial leadership in nanotechnology poses a variety of technical and policy challenges, including development of technologies that will enable commercial scale manufacturing of nanotechnology materials and products; environmental, health, and safety concerns; and maintenance of public confidence in its safety.

Congress established programs, assigned responsibilities, and initiated research and development related to these issues in the 21st Century Nanotechnology Research and Development Act of 2003 (P.L. 108-153). Although many provisions of this act have no sunset provision, FY2008 was the last year of agency authorizations included in the act. Legislation to amend and reauthorize the act was introduced in the House (H.R. 5940, 110th Congress) and the Senate (S. 3274, 110th Congress) in the 110th Congress. The House passed H.R. 5940 by a vote of 407-6; the Senate did not act on S. 3274. In January 2009, H.R. 554 (111th Congress), the National Nanotechnology Initiative Amendments Act of 2009, was introduced in the 111th Congress. The act contained essentially the same provisions as H.R. 5940. In February 2009, the House passed the bill by voice vote under a suspension of the rules. The bill was referred to the Senate Committee on Commerce, Science, and Transportation; no further action was taken. On May 7, 2010, the House Committee on Science and Technology reported the America COMPETES Reauthorization Act of 2010 (H.R. 5116, 111th Congress) which included, as Title I, Subtitle A, of the National Nanotechnology Initiative Amendments Act of 2010. This title was removed prior to enactment. No comprehensive reauthorization bill has been introduced in the 112th Congress.

Proponents of the NNI assert that nanotechnology is one of the most important emerging and enabling technologies and that U.S. competitiveness, technological leadership, national security, and societal interests require an aggressive approach to its development and commercialization. Critics of the NNI voice concerns that reflect disparate underlying beliefs. Some critics assert that the government is not doing enough to move technology from the laboratory into the marketplace. Others argue that the magnitude of the public investment may skew what should be market-based decisions in research, development, and commercialization. Still other critics say that the inherent risks of nanotechnology are not being addressed in a timely or effective manner.

Introduction

Nanotechnology has been an issue of interest to Congress for a number of years, coming into focus in 2000 with the launch of the U.S. National Nanotechnology Initiative (NNI) by President Clinton in his FY2001 budget request to Congress. From FY2001 through FY2012, Congress appropriated approximately $15.6 billion for nanotechnology research and development (R&D). President Obama has proposed $1.8 billion in NNI funding for FY2013. The NNI's efforts have been directed at advancing understanding and control of matter at the nanoscale,[1] where the physical, chemical, and biological properties of materials differ in fundamental and useful ways from the properties of individual atoms or bulk matter.[2]

Nanotechnology: A Description

The term "nanotechnology" is often used as an all-encompassing term for nanoscale science, engineering, and technology. Nanotechnology is the understanding and control of matter at dimensions of roughly 1 to 100 nanometers, the size-scale between individual atoms and bulk materials, where unique phenomena enable novel applications. A nanometer is one-billionth of a meter, or about the width of 10 hydrogen atoms arranged side-by-side in a line. Nanotechnology involves imaging, measuring, modeling, and manipulating matter at this size-scale.

At the nanoscale, the physical, chemical, and biological properties of materials can differ in fundamental and useful ways from the properties of individual atoms and molecules or bulk matter. Nanotechnology R&D is directed toward understanding and creating improved materials, devices, and systems that exploit these new properties.

Physicist Richard Feynman's remarks at the 1959 annual meeting of the American Physical Society are often cited as the first articulation of and vision for nanotechnology. Though he did not use the term nanotechnology in this speech, he spoke of controlling matter at the nanoscale and creating atomic-level machines, positing some of the applications that doing so might enable.

Source: *The National Nanotechnology Initiative Strategic Plan, 2004*, Nanoscale Science, Engineering, and Technology Subcommittee, National Science and Technology Council, Executive Office of the President, December 2004.

The development and application of nanotechnology—more fully explained below—across a wide array of products and industries holds the potential for significant economic and societal benefits. To capture these benefits, the United States will have to effectively address a variety of technical and policy challenges that stand as potential barriers to commercialization, including environmental, health, and safety (EHS) concerns and their implications for workplace, environmental, food, and drug regulations; development of standards, reference materials, and consistent nomenclature; development of new measurement methods and tools; effective technology transfer to the private sector; protection of intellectual property; availability, affordability, and patience of investment capital; ethical, legal, and societal concerns; public understanding, support, and acceptance; and development of a world-class scientific and technical nanotechnology workforce.

This report provides an overview of nanotechnology, the National Nanotechnology Initiative, possible reauthorization of the 21st Century Nanotechnology Research and Development Act of 2003 (P.L. 108-153), and appropriations issues.

---

[1] In the context of the NNI and nanotechnology, the nanoscale refers to a dimension of 1 to 100 nanometers (see box on this page).

[2] While extensive R&D has been, and continues to be, conducted to understand and harness the properties of individual atoms, this is not the domain of nanotechnology.

Overview

The economic and societal promise of nanotechnology has led to involvement and investments by governments and companies around the world. In 2000, the United States became the first nation to establish a formal, national initiative to advance nanoscale science, engineering, and technology—the National Nanotechnology Initiative. Since then, the United States has emerged as a global leader in nanotechnology. However, the competition for global leadership is intensifying as foreign investments in nanoscale science, engineering, and technology increase. Other nations have followed the U.S. lead and established their own national nanotechnology programs, each with varying degrees of investment, foci, and support for industrial applications and commercialization. Today, almost every nation that supports research and development (R&D) has a national-level nanotechnology program.

In 2011, Lux Research, an emerging technologies consulting firm, estimated total (public and private) global nanotechnology funding for 2010 to be approximately $17.8 billion with corporate R&D ($9.6 billion) accounting for a majority of funding for the first time.[3] Cientifica, a privately held nanotechnology business analysis and consulting firm, estimated global public investments in nanotechnology in 2010 to be approximately $10 billion per year, with cumulative global public investments through 2011 reaching approximately $67.5 billion. Cientifica also concluded that the United States has fallen behind both Russia and China in nanotechnology R&D funding on a purchasing power parity (PPP) basis (which takes into account the price of goods and services in each nation), but still leads the world in real dollar terms (adjusted on a currency exchange rate basis).[4]

Global investments in nanotechnology already have begun to yield economic benefits as products incorporating nanotechnology enter the marketplace. These products are estimated to have produced $200 billion in revenues in 2008, including $80 billion in the United States.[5] By tapping the unique properties that emerge at the nanoscale, proponents maintain that nanotechnology holds the potential for products that could transform existing industries and create new ones, clean and protect the environment, extend and improve the quality of our lives, and strengthen national security. Most nanotechnology products currently on the market—such as faster computer processors, higher density memory devices, lighter-weight auto parts, stain-resistant clothing, antibiotic bandages, cosmetics, and clear sunscreen—are evolutionary in nature, offering incremental improvements in characteristics such as performance, aesthetics, cost, size, and weight.

Evolutionary nanotechnology products, however, represent only a small fraction of what many see as the substantial longer-term economic and societal promise of nanotechnology. One estimate projects nanotechnology product revenues will reach $3.1 trillion by 2015,[6] while

---

[3] Working Party on Nanotechnology, Organization for Economic Cooperation and Development, *OECD /NNI International Symposium on Assessing the Economic Impact of Nanotechnology, Background Paper 2: Finance and Investor Models in Nanotechnology,* March 16, 2012, p. 4.

[4] Cientifica, *Global Funding of Nanotechnologies and Its Impact,* July 2011, http://cientifica.eu/blog/wp-content/uploads/downloads/2011/07/Global-Nanotechnology-Funding-Report-2011.pdf.

[5] Mihail C. Roco, Chad A. Mirkin, and Mark C. Hersam, "The Long View of Nanotechnology Development: The NNI at Ten Years," in *Nanotechnology Research Directions for Societal Needs in 2020: Retrospective and Outlook* (Springer, 2011), p. 4.

[6] Lux Research, "Overhyped Technology Starts to Reach Potential," press release, July 22, 2008, (continued...)

---

another estimate projects revenues will reach $2.95 trillion by 2015, of which almost half will come from semiconductors.[7, 8]

Many nanotechnology advocates—including business executives, scientists, engineers, medical professionals, and venture capitalists—assert that in the longer term, nanotechnology, especially in combination with information technology, biotechnology, and the cognitive sciences, may deliver revolutionary advances, including:

- new prevention, detection, and treatment technologies that could reduce substantially death and suffering from cancer and other deadly illnesses;[9]

- new organs to replace damaged or diseased ones;[10]

- contact lenses, skin patches, and glucose-sensing tattoos that monitor diabetics' blood sugar levels and warn when too high or low;[11]

- clothing that protects against toxins and pathogens;[12]

- clean, inexpensive, renewable power through energy creation, storage, and transmission technologies;[13]

- inexpensive, portable water purification systems that provide universal access to safe water;[14]

- energy efficient, low-emission "green" manufacturing systems;[15]

- high-density memory systems capable of storing the entire Library of Congress collection on a device the size of a sugar cube;[16]

- agricultural technologies that increase crop yield and improve nutritional value, reducing global hunger and malnutrition;[17]

(...continued)

http://www.luxresearchinc.com/press/RELEASE_Nano-SMR_7_22_08.pdf.

[7] Cientifica, *Halfway to the Trillion Dollar Market: A Critical Review of the Diffusion of Nanotechnologies*, 2007, http://www.cientifica.eu/files/Whitepapers/A%20Reassessment%20of%20the%20Trillion%20WP.pdf.

[8] While views vary on how to calculate nanotechnology's contribution to these products, the consensus is that nanotechnology is likely to have a significant economic impact and transformative effect on many industries.

[9] National Cancer Institute website. http://nano.cancer.gov/resource_center/tech_backgrounder.asp

[10] Ibid.

[11] Aslan, Kadir; Lakowicz, Joseph R.; and Geddes, Chris D. "Nanogold plasmon resonance-based glucose sensing. Wavelength-ratiometric resonance light scattering," *Analytical Chemistry*, 2005, Vol. 77. National Institute of Diabetes and Digestive and Kidney Disease, National Institutes of Health, Department of Health and Human Services, *Strategic Plan for Pediatric Urology*,, February 2006.

[12] Risbud, Aditi. "Fruit of the Nano Loom," *Technology Review*, February 2006.

[13] Nanoscale Science, Engineering, and Technology (NSET) Subcommittee, National Science and Technology Council (NSTC), Executive Office of the President (EOP), *Nanoscience Research for Energy Needs*, December 2004.

[14] Risbud, Aditi. "Cheap Drinking Water from the Ocean," *Technology Review*, June 2006.

[15] Selko, Adrienne. "New Nanotechnology-Based Coatings Are Energy Efficient and Environmentally Sound," *Industry Week*, August 22, 2007. "Tomorrow's Green Nanofactories," *Science Daily*, July 11, 2007.

[16] Interagency Working Group on Nanoscience, Engineering, and Technology, NSTC, EOP, *National Nanotechnology Initiative—Leading to the Next Industrial Revolution*, http://www.ostp.gov/NSTC/html/iwgn/iwgn.fy01budsuppl/nni.pdf.

[17] U.S. Department of Agriculture, *21st Century Agriculture: A Critical Role for Science and Technology*, June 2003; (continued...)

- self-repairing materials;[18]

- powerful, small, inexpensive sensors that can warn of minute levels of toxins and pathogens in air, soil, or water;[19] and

- decontaminated industrial sites through environmental remediation.[20]

While some applications of nanotechnology have proven market-ready, much fundamental research remains ahead, including efforts to advance understanding of nanoscale phenomena; characterize nanoscale materials; understand how to control and manipulate nanoscale particles; develop instrumentation and measurement methods; and understand how nanoscale particles interact with humans, animals, plants, and the environment. In addition, several federal agencies—such as the Departments of Defense, Energy, and Homeland Security—see the potential for nanotechnology to help address mission requirements. Historically, the federal government has played a central role in funding these types of research and development activities.

Though federal nanoscale science, engineering, and technology R&D had been underway for over a decade, the NNI was first initiated as a Presidential technology initiative in 2000.[21] The original participating agencies were the National Science Foundation (NSF), the Department of Defense (DOD), the Department of Energy (DOE), the Department of Commerce's (DOC) National Institute of Standards and Technology (NIST), the National Aeronautics and Space Administration (NASA), and the Department of Health and Human Services' National Institutes of Health (NIH). In 2012, 26 agencies participate in the NNI, including 15 that have received appropriations to conduct and/or fund nanotechnology R&D. Since its first year of funding in FY2001, the NNI's annual appropriations grew nearly four-fold to an estimated $1.7 billion in FY2012. President Obama has requested $1.8 billion in NNI funding for FY2013.[22]

In 2003, Congress provided a statutory foundation for some of the activities of the NNI through the 21st Century Nanotechnology Research and Development Act of 2003 (P.L. 108-153). The act established a National Nanotechnology Program (NNP) and provided authorizations for a subset of the NNI agencies, namely the NSF, DOE, NASA, NIST, and Environmental Protection Agency (EPA).[23] The act, however, did not address the participation of several agencies that fund

---

(...continued)

and *Nanoscale Science and Engineering for Agriculture and Food Systems: Draft Report of the National Planning Workshop to the Cooperative State Research, Education, and Extension Service of the U.S. Department of Agriculture,* July 2003.

[18] NSET Subcommittee, NSTC, EOP, *Nanotechnology in Space Exploration,* August 2004, http://www.nano.gov/nni_space_exploration_rpt.pdf.

[19] NSET Subcommittee, NSTC, EOP, *Nanotechnology and the Environment,* May 2003, http://www.nano.gov/NNI_Nanotechnology_and_the_Environment.pdf.

[20] U.S. Environmental Protection Agency, *Proceedings of the U.S. Environmental Protection Agency Workshop on Nanotechnology for Site Remediation,* October 2005.

[21] The White House, "National Nanotechnology Initiative: Leading to the Next Industrial Revolution," press release, January 21, 2000, http://clinton4.nara.gov/WH/New/html/20000121_4 html; and "Steering the technology that will redefine life as we know it," *Industrial Biotechnology,* Vol. 1, No. 3, Fall 2005, http://www.nsf.gov/crssprgm/nano/reports/mcr_ind_biotech_interview.pdf.

[22] NSET Subcommittee, NSTC, EOP, *The National Nanotechnology Initiative: Supplement to the President's FY2013 Budget,* February 2012.

[23] While many provisions of this act have no sunset provision, FY2008 was the last year of agency authorizations included in the act.

nanotechnology R&D under the NNI, including DOD, NIH, and the Department of Homeland Security (DHS). Nevertheless, coordination of nanotechnology R&D activities across all NNI funding agencies continues under the National Science and Technology Council's (NSTC's) Nanoscale Science, Engineering, and Technology (NSET) Subcommittee.[24] According to the NSET Subcommittee's 2004 NNI Strategic Plan, "For continuity and to capture this broader participation, the coordinated federal activities as a whole will continue to be referred to as the National Nanotechnology Initiative." Accordingly, the functions and activities established under the act are incorporated into the executive branch's implementation of the NNI.

The thrust of the NNI has primarily been the development of fundamental scientific knowledge through basic research. Investments at mission agencies, such as DOD, have supported nanotechnology applications development for which they are a primary customer. Other investments have supported infrastructural technologies. For example, NIST has contributed to developing tools and standards that enable measurement and control of matter at the nanoscale, thereby supporting the conduct of R&D and the ability to manufacture nanoscale materials and products. As understanding of nanotechnology has matured, the NNI has worked with a variety of industry organizations to facilitate the movement of research results from the laboratory bench to the marketplace in fields as disparate as semiconductors, chemicals, energy, concrete, and forest products.

The NNI agencies also have begun to address research needs and regulatory issues related to environmental, health, and safety, as well as issues such as public understanding and workforce education and training. The NNI agencies actively engage in a variety of international fora, such as the Organization for Economic Cooperation and Development (OECD) and the International Standards Organization (ISO), to cooperatively address nanotechnology issues related to EHS, metrology[25] and standards, nomenclature, and nanoscale materials characterization.

Maintaining U.S. leadership poses a variety of technical, economic, and policy challenges, including:

- safeguarding the environment and ensuring human health and safety;

- creating the standards, reference materials, nomenclature, methods, and tools for metrology to enable the manufacturing of nanoscale materials and products;

- developing a world-class scientific and technical nanotechnology workforce;

- translating research results into products, including effective technology transfer to the private sector;

- understanding public perceptions and attitudes and fostering public understanding;

- addressing ethical, legal, and societal implications;

- protecting intellectual property;

---

[24] Prior to P.L. 108-153, the Bob Stump Defense Authorization Act for Fiscal Year 2003 (P.L. 107-314) required DOD to "provide for interagency cooperation and collaboration on nanoscale research and development." The NSET Subcommittee is a subcommittee of the NSTC Committee on Technology.

[25] Metrology is the science of measurement, including the equipment and processes used to produce a measurement.

- securing investment capital for early-stage research, development, and commercialization; and

- fostering and facilitating international cooperation and coordination.

Proponents of the NNI assert that nanotechnology is one of the most important emerging and enabling technologies[26] and that U.S. competitiveness, technological leadership, national security, and societal interests require an aggressive approach to the development and commercialization of nanotechnology. Critics of the NNI hold a variety of competing views, asserting that government is not doing enough, is doing too much, or is moving too quickly.

Some in industry have criticized the NNI for being overly focused on basic research and not being aggressive enough in moving NNI-funded R&D out of government and university laboratories and into industry. Others in industry have criticized the federal government for not providing mechanisms to help advance nanotechnology R&D to the point where it becomes economically viable for venture capitalists, corporations, and other investors to create products and bring them to market. Some refer to this gap as the "valley of death."[27] Still others in industry have criticized the NNI for not adequately supporting the development of metrology, standards, equipment, and processes necessary to manufacture nanotechnology materials, products, and systems at a commercial scale.

Conversely, supporters of industry-driven market investments contend that extensive government support for nanotechnology may supplant the judgment of the marketplace by picking "winners and losers" in technological development. For example, the size and directions of the NNI investments may encourage industry to follow the government's lead rather than independently selecting R&D directions itself or, alternatively, may result in the promotion of a less effective technology path over a more effective one. These supporters also assert that federal government funding of scientific research is often wasteful, driven by political considerations and not scientific merit.[28]

Some non-governmental organizations (NGOs) are critical of nanotechnology for its potential adverse impacts on human health and safety and on the environment. They assert that the government is pushing ahead too quickly in developing nanotechnology and encouraging its commercialization and use without adequately investing in research focused on understanding and mitigating negative EHS implications.[29] They argue that the very characteristics that make nanotechnology promising also present significant potential risks to human health and safety and the environment. Some of these critics argue for application of the "precautionary principle,"

---

[26] The Department of Commerce has characterized emerging and enabling technologies as those that "offer a wide breadth of potential application and form an important technical basis for future commercial applications." (ATP Rule, 15 C.F.R. Part 295).

[27] The term "valley of death" is used by business executives, economists, and venture capitalists to describe the development gap that often exists between a laboratory discovery and the market's willingness to invest to advance the discovery to a final commercial product. This gap occurs due to a variety of issues, such as technical risk, market uncertainty, and likelihood of obtaining an adequate return on investment.

[28] Crews, Clyde Wayne, Jr., "Washington's Big Little Pork Barrel: Nanotechnology," Cato Institute website, May 29, 2003.

[29] Testimony of Andrew Maynard, Chief Science Advisor, Project on Emerging Nanotechnologies, Woodrow Wilson International Center for Scholars, "Research on Environmental and Safety Impacts of Nanotechnology: Current Status of Planning and Implementation under the National Nanotechnology Initiative," hearing, Subcommittee on Research and Science Education, House Committee on Science and Technology, October 31, 2007.

which holds that regulatory action may be required to control potentially hazardous substances even before a causal link has been established by scientific evidence.[30] At least one NGO has called for a moratorium on nanotechnology R&D and new commercial products incorporating synthetic nanoparticles.[31]

National Nanotechnology Initiative

The National Nanotechnology Initiative is an interagency program that coordinates federal nanoscale science, engineering, and technology R&D activities and related efforts among participating agencies.

Vision and Goals

The National Science and Technology Council has stated the following vision for the NNI:

> The vision of the NNI is a future in which the ability to understand and control matter at the nanoscale leads to a revolution in technology and industry that benefits society. The NNI expedites the discovery, development, and deployment of nanoscale science, engineering, and technology to serve the public good, through a program of coordinated research and development aligned with the missions of the participating agencies.[32]

To achieve its vision, the NNI has established four goals:

- advance a world-class R&D program;

- foster the transfer of new technologies into products for commercial and public benefit;

- develop and sustain educational resources, a skilled workforce, and the supporting infrastructure and tools to advance nanotechnology; and

- support responsible development of nanotechnology.[33]

---

[30] "NGOs urge precautionary principle in use of nanomaterials," EurActiv.com, June 14, 2007. http://www.euractiv.com/en/environment/ngos-urge-precautionary-principle-use-nanomaterials/article-164619 Sass, Jennifer. "Nanotechnology and the Precautionary Principle," presentation, Natural Resources Defense Council, 2006. http://docs nrdc.org/health/hea_06121402a.pdf The precautionary principle has been used in other countries on some issues. For example, the Biosafety Protocol to the 1992 Convention on Biological Diversity incorporates provisions applying the precautionary principle to the safe handling, transfer, and trade of genetically modified organisms. For further information, see CRS Report RL30594, *Biosafety Protocol for Genetically Modified Organisms: Overview*, by Alejandro E. Segarra and Susan R. Fletcher.

[31] ETC Group, "No Small Matter II: The Case for a Global Moratorium—Size Matters!," Occasional Paper Series, April 2003, http://www.etcgroup.org/upload/publication/pdf_file/165.

[32] NSET Subcommittee, NSTC, EOP, *The National Nanotechnology Initiative Strategic Plan,* February 2011, http://www.nano.gov/sites/default/files/pub_resource/2011_strategic_plan.pdf.

[33] Ibid.

History

Attempts to coordinate federal nanoscale R&D began in November 1996, as staff members from several agencies met regularly to discuss their plans and programs in nanoscale science and technology. This group continued informally until September 1998, when it was designated as the Interagency Working Group on Nanotechnology (IWGN) under the NSTC. In August 1999, IWGN completed its first draft of a plan for an initiative in nanoscale science and technology, which was subsequently approved by the President's Council of Advisors on Science and Technology (PCAST) and the White House Office of Science and Technology Policy (OSTP).[34]

In his 2001 budget submission to Congress, then-President Clinton raised nanotechnology-related research to the level of a federal initiative, officially referring to it as the National Nanotechnology Initiative.[35]

Legislative Approach

Congress has played a central role in the National Nanotechnology Initiative, providing appropriations for the conduct of nanoscale science, engineering, and technology research; establishing programs; and creating a legislative foundation for the activities of the NNI.

Congressional funding for the NNI is provided through appropriations to each of the NNI-participating agencies. The NNI has no centralized funding. The overall NNI budget is calculated by aggregating the nanotechnology budgets for each of the federal agencies that conduct or provide funding for nanoscale R&D.

In FY2001, the first year of NNI funding, Congress provided $464 million to eight agencies for nanoscale R&D.[36] The NNI has continued to receive support from both Congress and the White House. Both the number of agencies participating in the NNI and the size of the federal investment have grown. Currently 26 agencies participate in the NNI, 15 of which have received appropriated funds for nanotechnology R&D.[37] Total NNI funding in FY2012 is approximately $1.7 billion,. The original six agencies identified at the launch of the NNI[38] still account for the vast majority of NNI funding (96.1% in FY2012).

---

[34] National Nanotechnology Initiative website, http://www nano.gov/html/about/history.html.

[35] The White House, "National Nanotechnology Initiative: Leading to the Next Industrial Revolution," press release, January 21, 2000. http://clinton4.nara.gov/WH/New/html/20000121_4 html; and National Nanotechnology Initiative website, http://www nano.gov/html/about/history.html.

[36] In its January 21, 2001 press release, "National Nanotechnology Initiative: Leading to the Next Industrial Revolution," announcing the establishment of the NNI, the White House identified only six participating agencies— NSF, DOD, DOE, NIST, NASA, and NIH. Subsequently, EPA and DOJ reported nanotechnology R&D funding in FY2001, bringing the total number of agencies funding nanotechnology R&D in FY2001 to eight.

[37] NNI participants include agencies that either conduct or provide funding for nanotechnology R&D, as well as agencies with missions that may affect the development, commercialization, and use of nanotechnology. For example, in the latter case, the Food and Drug Administration may regulate (or not regulate) nanotechnology products, the U.S. Patent and Trademark Office's (USPTO) treatment of nanotechnology-related patents may affect the value of the underlying intellectual property, and the execution of the missions of the Departments of Education and Labor could affect the preparedness of the U.S. workforce for emerging nanotechnology jobs. Some nanotechnology R&D agencies may also have non-R&D missions related to nanotechnology. For example, EPA conducts and funds R&D but also has a regulatory mission that could affect nanotechnology research, development, production, use, and/or disposal.

[38] The original six agencies identified at the launch of the NNI were the Department of Defense, Department of Energy, (continued...)

21st Century Nanotechnology Research and Development Act of 2003

Congress codified and further defined some of the NNI's activities in the 21st Century Nanotechnology Research and Development Act of 2003 which was passed by Congress in November 2003, and signed into law (P.L. 108-153) by President Bush on December 3, 2003.[39] The legislation received strong bipartisan support in both the House of Representatives, which passed the bill on a recorded vote of 405-19, and in the Senate, which passed the bill by unanimous consent.

Though this act is often referred to as the enabling legislation for the National Nanotechnology Initiative, the act actually establishes a National Nanotechnology Program (NNP). The act provides authorizations for five NNI agencies—the National Science Foundation, Department of Energy, NASA, National Institute of Standards and Technology, and Environmental Protection Agency—but not for the Department of Defense, National Institutes of Health, Department of Homeland Security,[40] or other NNI research agencies that collectively accounted for 46% of NNI funding in FY2003.

The act created the NNP for the purposes of establishing the goals, priorities, and metrics for evaluation of federal nanotechnology research, development, and other activities; investing in federal R&D programs in nanotechnology and related sciences to achieve those goals; and providing for interagency coordination of federal nanotechnology research, development, and other activities undertaken pursuant to the NNP.

Key provisions of the act include:

- authorizing appropriations for the nanotechnology-related activities of the NSF, DOE, NASA, NIST, and EPA for fiscal years 2005 through 2008, totaling $3.679 billion for the four year period;

- establishing a National Nanotechnology Coordination Office, with a director and full time staff to provide administrative support to the NSTC;

- establishing a National Nanotechnology Advisory Panel (NNAP) to advise the President and the NSTC on matters relating to the NNP;

- establishing a triennial review of the NNP by the National Research Council of the National Academy of Sciences;

(...continued)

National Institute of Standards and Technology (Department of Commerce), National Science Foundation, National Aeronautics and Space Administration, and National Institutes of Health (DHHS). The White House, "National Nanotechnology Initiative: Leading to the Next Industrial Revolution," press release, January 21, 2000. http://clinton4 nara.gov/WH/New/html/20000121_4 html; and National Nanotechnology Initiative website. http://www nano.gov/html/about/history.html.

[39] U.S. Congress. 2003. 21st Century Nanotechnology Research and Development Act. P.L. 108-153. 15 U.S.C. 7501. 108 Cong., December 3.

[40] FY2003 funding attributed to DHS for the purpose of this calculation is based on nanotechnology R&D appropriations received by the Department of Transportation's Transportation Security Administration (TSA). TSA was transferred to DHS in the Homeland Security Act of 2002 (P.L. 107-296) which was enacted after the start of FY2003.

- directing the NSTC to oversee the planning, management, and coordination of the program, including the development of a triennial strategic plan;

- directing the Department of Commerce's National Institute of Standards and Technology to establish a program to conduct basic research on issues related to the development and manufacture of nanotechnology, and to use the Manufacturing Extension Partnership program to ensure results reach small- and medium-sized manufacturing companies;

- directing the Secretary of Commerce to use the National Technical Information Service to establish a clearinghouse of information related to commercialization of nanotechnology research;

- directing the Secretary of Energy to establish a program to support consortia to conduct interdisciplinary nanotechnology R&D designed to integrate newly developed nanotechnology and microfluidic tools with systems biology and molecular imaging;

- directing the Secretary of Energy to carry out projects to develop, plan, construct, acquire, operate, or support special equipment, instrumentation, or facilities for investigators conducting nanotechnology R&D; and

- directing the establishment of two centers, on a merit-reviewed and competitive basis: (1) the American Nanotechnology Preparedness Center, to conduct, coordinate, collect, and disseminate studies on the societal, ethical, environmental, educational, legal, and workforce implications of nanotechnology; and to identify anticipated issues related to the responsible research, development, and application of nanotechnology, as well as provide recommendations for preventing or addressing such issues; and (2) the Center for Nanomaterials Manufacturing, to encourage, conduct, coordinate, commission, collect, and disseminate research on new manufacturing technologies for materials, devices, and systems with new combinations of characteristics, such as, but not limited to, strength, toughness, density, conductivity, flame resistance, and membrane separation characteristics; and to develop mechanisms to transfer such manufacturing technologies to U.S. industries.

While the act establishes a National Nanotechnology Program, the executive branch continues its broader effort under the NNI framework and name. According to the NNI's 2004 Strategic Plan:

> Many of the activities outlined in the Act were already in progress as part of the NNI. Moreover, the ongoing management of the initiative involves considerable input from Federal agencies that are not named specifically in the Act.... For continuity, and to capture this broader participation, the coordinated Federal activities as a whole will continue to be referred to as the National Nanotechnology Initiative.[41]

Reauthorization Efforts

The 21st Century Nanotechnology Research and Development Act provided a legislative foundation for some of the activities of the NNI, authorized agency funding levels through

---

[41] NSET Subcommittee, NSTC, EOP, *The National Nanotechnology Initiative Strategic Plan*, December 2004, http://www nano.gov/NNI_Strategic_Plan_2004.pdf.

FY2008, and sought to address challenges associated with the development and commercialization of nanotechnology. While many provisions of this act have no sunset provision, FY2008 was the last year of agency authorizations included in the act.

Legislation to amend and reauthorize the act was introduced in the House in both the 110th Congress and 111th Congress:

- H.R. 5940 (110th Congress) and S. 3274 (110th Congress) were both titled the National Nanotechnology Initiative Amendments Act of 2008. The House passed H.R. 5940 by a vote of 407-6; the Senate did not act on S. 3274.

- H.R. 554 (111th Congress), the National Nanotechnology Initiative Amendments Act of 2009, contained essentially the same provisions as H.R. 5940 (110th Congress). In February 2009, the House passed the bill by voice vote under a suspension of the rules. The Senate did not act on H.R. 554.

- S. 1482 (111th Congress), the National Nanotechnology Initiative Amendments Act of 2009, was introduced in the Senate and referred to the Committee on Commerce, Science, and Transportation. No further action was taken.

- H.R. 820 (111th Congress), the Nanotechnology Advancement and New Opportunities Act, also would have amended P.L. 108-153. The provisions of H.R. 820 covered a variety of jurisdictions, thus the bill was assigned to multiple House committees. No further action was taken.

- On May 7, 2010, the House Committee on Science and Technology reported the America COMPETES Reauthorization Act of 2010 (H.R. 5116, 111th Congress) which included, as Title I, Subtitle A, the "National Nanotechnology Initiative Amendments Act of 2010." Provisions of this subtitle were nearly identical to the provisions of H.R. 554 (111th Congress). This title was removed from the bill prior to its enactment.

Although H.R. 2749, the Nanotechnology Advancement and New Opportunities Act, would amend the 21st Century Nanotechnology Research and Development Act, no comprehensive reauthorization legislation has been introduced in the 112th Congress. For additional information, see "Selected Nanotechnology Legislation in the 111th and 112th Congress."

Structure

Nanoscale Science, Engineering, and Technology (NSET) Subcommittee

The NNI is coordinated within the White House through the NSTC, the Cabinet-level council by which the President coordinates science, space, and technology policies across the federal government. Operationally, NNI coordination is accomplished through the Nanoscale Science, Engineering, and Technology Subcommittee of the NSTC's Committee on Technology (CT). The NSET Subcommittee also has an informal reporting relationship to the NSTC's Committee on Science (CS). The NSET Subcommittee is led by an agency co-chair, currently from DOD, and an OSTP co-chair. The NSET Subcommittee is comprised of representatives from 26 federal entities (including 15 that have funded, over the course of the NNI, nanotechnology R&D), OSTP and the Office of Management and Budget.[42]

The NSET Subcommittee has established several working groups, each taking on efforts in key subject areas.[43] Among them:

National Environmental and Health Implications (NEHI)

The National Environmental and Health Implications (NEHI) working group was chartered to provide for exchange of information among agencies that support research and those responsible for regulations and guidelines related to nanotechnology products; to facilitate identification, prioritization, and implementation of research and other activities required for the responsible research, development, utilization, and oversight of nanotechnology; and to promote communication of information related to research on environmental and health implications of nanotechnology to other government agencies and non-government parties. To this end, the NEHI working group seeks to identify and prioritize EHS research needs related to nanotechnology. Sixteen NNI agencies (as well as OSTP and OMB) participate in the NEHI working group, including agencies that fund safety-related nanotechnology research and/or have regulatory authorities to guide the safe use of nanomaterials.

Nanomanufacturing, Industry Liaison, and Innovation(NILI)

The Nanomanufacturing, Industry Liaison, and Innovation (NILI) working group was chartered to enhance collaboration and information sharing between U.S. industry and government on nanotechnology-related activities to advance and accelerate the creation of new products and manufacturing processes derived from discovery at the nanoscale. It also facilitates federal, regional, state, and local nanotechnology R&D and commercialization activities. In addition, the NILI working group is to create innovative methods for transferring federally funded technology to industry. The NILI working group has facilitated collaborations between the NNI and the semiconductor/electronics industry, chemical industry, forest products industry, and the Industrial Research Institute.[44]

---

[42] The agencies that participate in the NSET Subcommittee comprise the NNI. NSET Subcommittee members include Forest Service, USDA; National Institute of Food and Agriculture, USDA; Agricultural Research Service, USDA; Bureau of Industry and Security, DOC; Consumer Product Safety Commission; DOD; Department of Education; DOE; DHS; Department of Justice; Department of Labor; Department of State; Department of Transportation; Department of the Treasury; Director of National Intelligence; EPA; Food and Drug Administration, HHS; International Trade Commission; NASA; NIH, Department of Health and Human Services (HHS); NIOSH, Center for Disease Control, HHS; NIST, DOC; NSF; Nuclear Regulatory Commission; U.S. Geological Survey, Department of the Interior; and U.S. Patent and Trademark Office, DOC. The Department of Commerce's Technology Administration was a participating agency in the NNI until its elimination in August 2007 under the America COMPETES Act (P.L. 110-69).

[43] NSET Subcommittee, NSTC, EOP, *The National Nanotechnology Initiative: Research and Development Leading to a Revolution in Technology and Industry-Supplement to the President's FY2008 Budget*, July 2007, http://www.nano.gov/NNI_08Budget.pdf.

[44] The Industrial Research Institute is an association of companies and federally funded laboratories with the mission of improving R&D capabilities through the development and dissemination of best practices.

Global Issues in Nanotechnology (GIN)

The Global Issues in Nanotechnology (GIN) working group was chartered to monitor foreign nanotechnology programs and development; broaden international collaboration on nanotechnology R&D, including safeguarding the environment and human health; and promote U.S. commercial and trade interests in nanotechnology. The NEHI working group works with the GIN working group to coordinate the U.S. position and participation in international activities related to EHS implications of nanotechnology. The GIN working group facilitates international collaboration on pre-competitive and non-competitive aspects of nanotechnology, and international engagement on trade, commercialization and regulatory issues. Fourteen NNI agencies participate in the GIN working group, as well as OSTP, OMB, and the Office of the U.S. Trade Representative (USTR).

Nanotechnology Public Engagement and Communications (NPEC)

The Nanotechnology Public Engagement and Communications (NPEC) working group was established to is "to encourage, coordinate, and support NNI member agencies and interagency efforts toward educating and engaging the public, policy makers, and stakeholder groups regarding nanotechnology, its applications and implications, and the work of the NNI."[45]

National Nanotechnology Coordination Office

The National Nanotechnology Coordination Office (NNCO) provides administrative and technical support to the NSET Subcommittee. Initially established in 2001 through a memorandum of understanding among the NNI participating agencies, the NNCO was authorized by the 21st Century Nanotechnology Research and Development Act of 2003 (P.L. 108-153). The NNCO was charged under the act with providing technical and administrative support to the NSTC and NNAP; serving as the point of contact for information on federal nanotechnology activities for the exchange of technical and programmatic information among stakeholders; conducting public outreach; and promoting access to and early application of NNP technologies, innovation, and expertise.

In addition, the NNCO serves as a liaison to academia, industry, professional societies, foreign organizations, and others facilitating the exchange of technical and programmatic information. The NNCO also coordinates preparation and publication of NNI interagency planning, budget, and assessment documents, and maintains the NNI website, http://www.nano.gov.

The act authorizes the work of the NNCO to be funded by contributions from NSET Subcommittee member agencies. According to the NNCO, funding is provided through a memorandum of understanding signed by eight NNI agencies. In principle, each agency contributes to the NNCO budget in proportion to its share of the President's total nanotechnology budget request for the signatory agencies. However, two of the signatories, EPA and DOT, had sufficiently small enough nanotechnology budgets in the early years of the NNI that they were not expected to contribute. EPA now contributes to funding the NNCO. Total NNCO funding from the agencies in FY2012 was approximately $3 million.

---

[45] NNI website, http://www.nano.gov/npec.

Figure 1. Organizations With a Role in the National Nanotechnology Initiative and Their Relationships

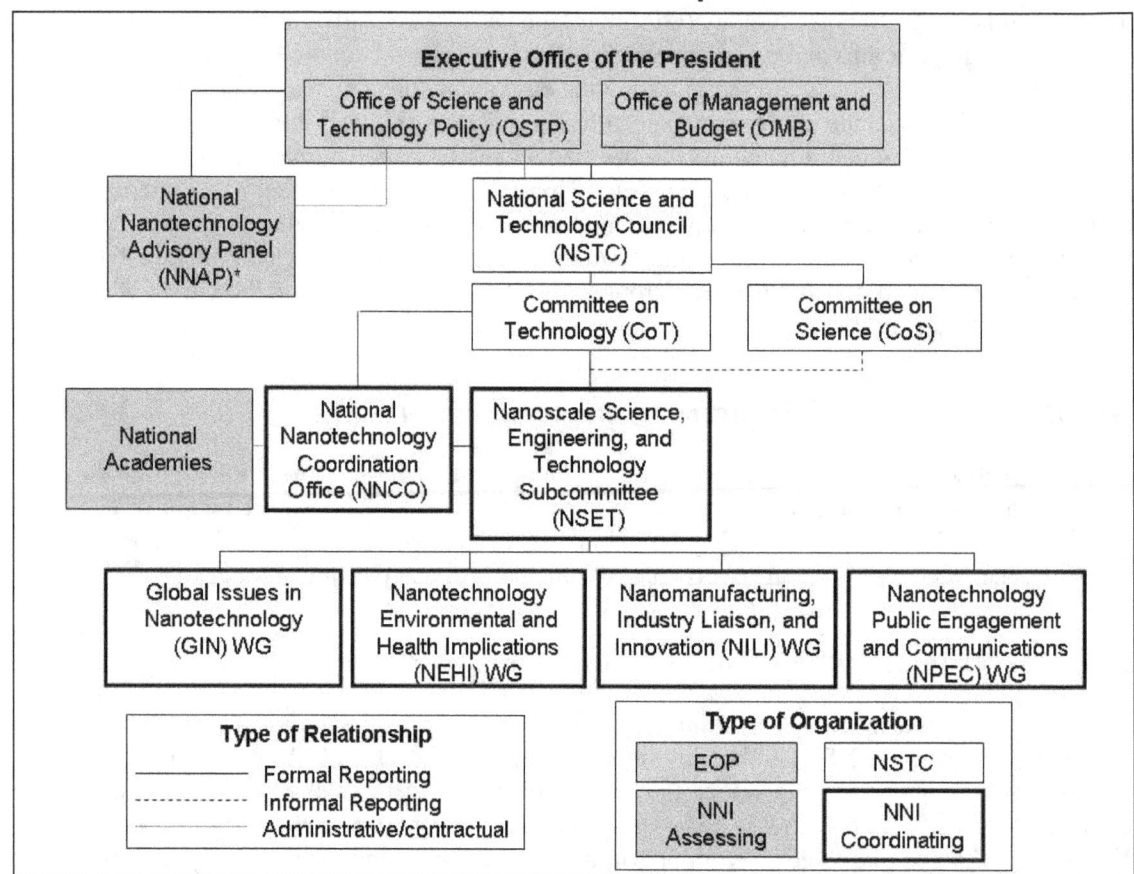

Source: Reproduced from *The National Nanotechnology Strategic Plan*, Nanoscale, Science, Engineering, and Technology Subcommittee, National Science and Technology Council, The White House, February 2011.

Note: Executive Order 13539 designates the President's Council of Advisors on Science and Technology (PCAST) as the National Nanotechnology Advisory Panel (NNAP), http://www.gpo.gov/fdsys/pkg/FR-2010-04-27/pdf/2010-9796.pdf.

Funding

The NNI supports fundamental and applied research on nanotechnology by funding research, creating multidisciplinary centers of excellence, and developing key research infrastructure. It also supports activities aimed at addressing the societal implications of nanotechnology, including ethical, legal, human and environmental health, and workforce issues.

This section provides information on NNI funding from two perspectives: organizationally by agency and functionally by program component area.

Agency Funding

The NNI budget is an aggregation of the nanotechnology components of the individual budgets of NNI-participating agencies. The NNI budget is not a single, centralized source of funds that is allocated to individual agencies. In fact, agency nanotechnology budgets are developed internally as part of each agency's overall budget development process. These budgets are subjected to review, revision, and approval by the Office of Management and Budget and become part of the President's annual budget submission to Congress. The NNI budget is then calculated by aggregating the nanotechnology components of the appropriations provided by Congress to each federal agency.

For FY2012, NNI R&D funding totals an estimated $1.697 billion, a $150.4 million (-8.1%) decrease from FY2011. The overall reduction was driven primarily by cuts in the R&D budgets of DOD (-$64.1 million, 15.1%), NSF (-$59.1 million, 12.2%), and DOE (-$30.8 million, -8.9%). The chronology of NNI funding by agency is detailed in **Table 1**.

President Obama requested $1.767 billion in funding for the NNI in FY2013, an increase of $70.1 million (4.1%). The FY2013 NNI budget request would support a broad range of programs among 16 agencies. Five agencies account for 96% of NNI funding in FY2012:[46]

- NSF (25.1%), which supports fundamental nanotechnology research across science and engineering disciplines;

- DOD (21.3%), whose investments in nanotechnology are aimed at addressing the department's national security mission;

- DOE (18.6%), which supports nanotechnology research providing a basis for new and improved energy efficiency, production, storage, and transmission technologies;

- NIH (25.4%), which emphasizes nanotechnology-based biomedical advances occurring at the intersection of biology and the physical sciences; and

- NIST (5.6%), which focuses on research in instrumentation, measurement, standards, characterization, and nanomanufacturing.

Other agencies investing in mission-related nanotechnology R&D are: NASA; EPA; the National Institute of Food and Agriculture (NIFA), Forest Service, and Agricultural Research Service (ARS) at the Department of Agriculture (USDA); the National Institute of Occupational Safety and Health (NIOSH) and Food and Drug Administration (FDA) at the Department of Health and Human Services (HHS); DHS; Department of Justice (DOJ); Department of Transportation's (DOT) Federal Highway Administration (FHWA); and the Consumer Product Safety Commission (CPSC).

---

[46] Based on FY2010 actual funding levels.

Table 1. NNI Funding, by Agency: FY2001-FY2012 and FY2013 Request

(in millions of current dollars)

Agency	FY 2001 Actual	FY 2002 Actual	FY 2003 Actual	FY 2004 Actual	FY 2005 Actual	FY 2006 Actual	FY 2007 Actual	FY 2008 Actual	FY 2009 Actual	ARRA (P.L. 111-5)[a]	FY2010 Actual	FY2011 Actual	FY2012 Estimate	FY2013 Request
Department of Energy	88	89	134	202	208	231	236	245	332.6	293.2	373.8	346.2	315.4	442.5
National Science Foundation	150	204	221	256	335	360	389	409	408.6	101.2	428.7	485.1	426.0	434.9
National Institutes of Health (HHS)	40	59	78	106	165	192	215	305	342.8	73.4	456.8	408.6	409.6	408.7
Department of Defense[b]	125	224	220	291	352	424	450	460	459.0		439.6	425.3	361.2	289.4
Nat'l Inst. of Standards and Tech. (DOC)	33	77	64	77	79	78	88	86	93.4	43.4	114.7	95.9	95.4	102.1
NASA	22	35	36	47	45	50	20	17	13.7		19.7	17.0	23.0	22.0
Environmental Protection Agency	5	6	5	5	7	5	8	12	11.6		17.7	17.4	17.5	19.3
Food and Drug Administration (HHS)									6.5		7.3	9.9	11.8	11.1
Nat'l Inst. for Occupational Safety (HHS)					3	4	7	7	6.7		8.5	10.0	10.0	10.0
Nat'l Inst. of Food and Agriculture (USDA)[c]			1	2	3	4	4	6	9.9		13.2	10.0	10.0	10.0
Department of Homeland Security		2		1	1	2	2	3	9.1		21.9	9.0	7.0	6.0
Forest Service (USDA)						2	3	5	5.4		7.1	10.0	5.0	5.0
Agricultural Research Service (USDA)													2.0	2.0
Federal Highway Administration (DOT)			1			1	1	1	0.9		3.2	1.0	1.0	2.0
Consumer Product Safety Commission									0.2		0.5	1.8	2.0	2.0
Department of Justice		1	1	2	2	<1	2	0	1.2		0.2			
TOTAL[d]	464	697	760	989	1,200	1,351	1,425	1,554	1,701.6	511.3	1,912.8	1,847.3	1,696.9	1767.0

Source: NNI website, http://www.nano.gov; NSET Subcommittee, NSTC, EOP, *The National Nanotechnology Initiative: Research and Development Leading to a Revolution in Technology and Industry, Supplement to the President's FY2013 Budget*, February 2012.

a. Funding figures for nanotechnology-related R&D under the ARRA are preliminary estimates.

b. According to NSTC, the DOD budgets shown above include congressionally directed funding of approximately $76 million in FY2006, $63 million in FY2007, $117 million in FY2009, and $75 million in FY2010. According to NSTC, the 2008 DOD estimate "includes many earmarks that are outside the NNI plan."

c. Formerly, the USDA Cooperative State Research, Education, and Extension Service (CSREES).

d. Totals may differ from the sum of the components due to rounding.

Program Component Area Funding

The 21st Century Nanotechnology R&D Act of 2003 called for the NSET Subcommittee to develop categories of investment called Program Component Areas (PCA) to provide a means by which Congress and the executive branch can be informed of and direct the relative investments in these areas. The PCAs are categories of investments that cut across the needs and interests of individual agencies and contribute to the achievement of one or more of the NNI's goals. The 2004 NNI strategic plan identified seven PCAs. The 2007 NNI strategic plan split the seventh PCA, Societal Dimensions, into two PCAs: Environment, Health, and Safety; and Education and Societal Dimensions. A description of the seven initial PCAs and their current funding is provided below,[47] as well as a description of the two derivative PCAs.[48] The chronology of NNI funding by PCA is detailed in **Table 2**.

In the following analysis of funding for each of the PCAs, FY2013 request levels are compared to FY2012 estimated levels.

Fundamental Phenomena and Processes

Fundamental Phenomena and Processes includes investments in the discovery and development of fundamental knowledge pertaining to the new phenomena in the physical, biological, and engineering sciences that occur at the nanoscale, as well as in understanding and articulation of scientific and engineering principles related to nanoscale structures, processes, and mechanisms.

Under the President's FY2013 request, funding for Fundamental Phenomena and Processes would decrease to $498.2 million (down $11.9 million, 2.3% from the FY2012 estimated level) due largely to a decrease in DOD funding of $24.0 million (-14.8%) which was partially offset by an increase in DOE funding of $13.4 million (12.9%).[49]

Nanomaterials

Nanomaterials includes research investments to discover novel nanoscale and nanostructured materials. This PCA also attempts to understand the properties of nanomaterials, and supports R&D to enable the design and synthesis, in a controlled manner, of nanoscale materials with targeted properties.

Under the President's FY2013 request, funding for Nanomaterials would rise to $368.4 million (up $57.7 million, 18.6% from the FY2012 estimated level), due almost entirely to an increase in DOE funding in this PCA (up $63.9 million, 77.6%). This increase would be offset, in part, by a proposed decrease in DOD funding (down $9.5 million, -22.6%).[50]

---

[47] NSET Subcommittee, NSTC, EOP, *The National Nanotechnology Initiative Strategic Plan,* December 2004, http://www.nano.gov/NNI_Strategic_Plan_2004.pdf.

[48] NSET Subcommittee, NSTC, EOP, *The National Nanotechnology Initiative Strategic Plan,* December 2007, http://www.nano.gov/NNI_Strategic_Plan_2007.pdf.

[49] NSET Subcommittee, NSTC, EOP, *The National Nanotechnology Initiative: Research and Development Leading to a Revolution in Technology and Industry, Supplement to the President's FY2013 Budget*, February 2012.

[50] Ibid.

Nanoscale Devices and Systems

Nanoscale Devices and Systems includes R&D investments that apply nanoscale science and engineering principles to create novel devices and systems or to improve existing ones. It also includes the use of nanoscale or nanostructured materials to achieve improved performance or new functionality. To meet this definition, the enabling science and technology must be at the nanoscale, but the systems and devices are not restricted to that size.

Under the President's FY2013 request, funding for Nanoscale Devices and Systems would remain essentially unchanged at $412.9 million (down $0.5 million, -0.1% from the FY2012 estimated level). An increase in DOE funding to $39.7 million (up $30.6 million, 335.4%) in this PCA would be offset by a decrease in DOD funding (down $31.2 million, -24.6).[51]

Instrumentation Research, Metrology, and Standards

The Instrumentation Research, Metrology, and Standards PCA includes R&D investments for development of tools needed to advance nanotechnology research and commercialization. Instrumentation for characterization, measurement, synthesis, and design of nanotechnology materials, structures, devices, and systems is funded through this PCA. R&D and other activities related to development of standards, including standards for nomenclature, materials, characterization, testing, and manufacture, are also in this PCA.

Under the President's FY2013 request, funding for Instrumentation Research, Metrology, and Standards would fall to $69.2 million (down $6.0 million, -8.0% from the FY2012 estimated level). Funding among agencies performing work in this PCA would remain largely unchanged.[52]

Nanomanufacturing

Nanomanufacturing R&D supports the development of scalable, reliable, cost-effective manufacturing of nanoscale materials, structures, devices, and systems. It also includes R&D and integration of ultra-miniaturized top-down processes and complex bottom-up processes.[53]

Under the President's FY2013 request, funding for Nanomanufacturing would rise to $88.9 million (up $15.4 million, 21.0% from the FY2012 estimated actual level) due to increases in funding for DOE (up $9.0 million), DOC (up $6.5 million, 70.2%), and NSF (up $5.0 million, 10.5%). These increases would be partially offset by a decrease in DOD funding (down $5.0 million, -45.5%).[54]

---

[51] Ibid.

[52] Ibid.

[53] Top-down processes are those that achieve design features by removing material from a larger block of material; bottom-up processes begin with smaller building blocks (atoms or molecules) and achieve design features by putting them together, possibly using self-assembly or nanoscale additive manufacturing.

[54] NSET Subcommittee, NSTC, EOP, *The National Nanotechnology Initiative: Research and Development Leading to a Revolution in Technology and Industry, Supplement to the President's FY2013 Budget*, February 2012.

Major Research Facilities and Instrumentation Acquisition

This PCA includes investments in the establishment and ongoing operations of user facilities and networks, the acquisition of major instrumentation, and other activities related to infrastructure for the conduct of nanoscale science, engineering, and technology R&D.

Under the President's FY2013 request, funding for Major Research Facilities and Instrumentation Acquisition would increase to $189.9 million (up $12.8 million, 7.2% over the FY2012 estimated level), due entirely to an increase of $12.8 million (12.2%) in DOE funding for this PCA.[55]

Societal Dimensions

The Societal Dimensions PCA includes investments in research and other activities that address the broad implications of nanotechnology to society. This includes assessing benefits and risks through research directed at environmental, health, and safety impacts of nanotechnology development; risk assessment of such impacts; education-related activities, such as development of materials for schools, undergraduate programs, technical training, and public outreach; and research directed at identifying and quantifying the broad implications of nanotechnology for society, including social, economic, workforce, educational, ethical, and legal implications.

Under the 2007 NNI Strategic Plan, the Societal Dimensions PCA was divided into two separate PCAs: Environment, Health, and Safety, and Education and Societal Dimensions. PCA reporting now uses an eight PCA taxonomy. NSTC retroactively reported FY2007 Societal Dimensions PCA spending in the new PCAs.[56] The NSET Subcommittee characterizes the new PCAs as follows:[57]

Environment, Health, and Safety

This PCA addresses research primarily directed at understanding the environmental, health, and safety impacts of nanotechnology development and corresponding risk assessment, risk management, and methods for risk mitigation.

Under the President's FY2013 request, funding for Environment, Health, and Safety would rise to $105.4 million (up $2.7 million, 2.6% from the FY2012 estimated level) due to increases in NIST and EPA funding.[58]

Education and Societal Dimensions

This PCA addresses education-related activities such as development of materials for schools, undergraduate programs, technical training, and public communication, including outreach and engagement. Such activities include research directed at identifying and quantifying the broad

---

[55] Ibid.

[56] NSET Subcommittee, NSTC, EOP, *The National Nanotechnology Initiative: FY2010 Budget & Highlights*, May 2009, http://www.nano.gov/NNI_2010_budget_supplement.pdf.

[57] Ibid.

[58] NSET Subcommittee, NSTC, EOP, *The National Nanotechnology Initiative: Research and Development Leading to a Revolution in Technology and Industry, Supplement to the President's FY2013 Budget*, February 2012.

implications of nanotechnology society, including social, economic, workforce, educational, ethical, and legal implications.

Under the President's FY2013 request, funding for Education and Societal Dimensions would be identical to the FY2012 estimated level of $34.2 million with no changes in any agency funding.[59]

---

[59] Ibid.

Table 2. NNI Funding, by Program Component Area, FY2006-FY2012 and FY2013 Request

(in millions of current dollars)

PCA	FY2006 Actual	FY2007 Actual	FY2008 Actual	FY2009 Actual	ARRA (P.L. 111-5)	FY2010 Actual	FY2011 Actual	FY2012 Estimate	FY2013 Request	Change, FY2013 Req. vs. 2012 Est. Dollars	Change, FY2013 Req. vs. 2012 Est. Percent
Fundamental Phenomena and Processes	455.9	480.6	478.5	479.2	130.6	490.5	567.9	510.1	498.2	-11.9	-2.3%
Nanomaterials	265.1	258.3	285.1	331.9	178.3	358.9	348.2	310.7	368.4	57.7	18.6%
Nanoscale Devices and Systems	319.6	344.7	372.7	435.2	68.0	542.1	454.7	413.4	412.9	-0.5	-0.1%
Instrumentation Research, Metrology, and Standards	51.0	52.5	69.0	90.8	12.4	89.4	73.2	75.3	69.2	-6.0	-8.0%
Nanomanufacturing	33.8	48.1	47.1	75.6	28.5	84.8	92.0	73.5	88.9	15.4	21.0%
Major Research Facilities and Instrumentation Acquisition	152.4	152.4	196.4	177.6	72.5	190.2	185.8	177.0	189.9	12.8	7.2%
Societal Dimensions	73.5										
- Environment, Health, and Safety		48.3	67.9	74.5	12.0	90.2	88.0	102.7	105.4	2.7	2.6%
- Education & Societal Dimensions		39.2	37.7	36.8	9.0	66.9	37.5	34.2	34.2	0	0%
TOTAL[a]	1,351.2	1,424.1	1,554.4	1,701.5	511.3	1,912.8	1,847.3	1,696.9	1,767.0	70.2	4.1%

Sources: NSET Subcommittee, NSTC, EOP, *The National Nanotechnology Initiative: Research and Development Leading to a Revolution in Technology and Industry, Supplement to the President's FY2008 Budget,* July 2007. NSET Subcommittee, NSTC, EOP, *National Nanotechnology Initiative: FY2009 Budget & Highlights,* February 2008. NSET Subcommittee, NSTC, EOP, *The National Nanotechnology Initiative: Research and Development Leading to a Revolution in Technology and Industry, Supplement to the President's FY2010 Budget,* May 2009. NSET Subcommittee, NSTC, EOP, *The National Nanotechnology Initiative: Research and Development Leading to a Revolution in Technology and Industry, Supplement to the President's FY2011 Budget,* The White House, February 2010. NSET Subcommittee, NSTC, EOP, *The National Nanotechnology Initiative: Research and Development Leading to a Revolution in Technology and Industry, Supplement to the President's FY2012 Budget,* February 2011. NSET Subcommittee, NSTC, EOP, *The National Nanotechnology Initiative: Research and Development Leading to a Revolution in Technology and Industry, Supplement to the President's FY2013 Budget,* February 2012.

a.   Totals may differ from the sum of the components due to rounding.

Centers, Networks, and User Facilities

A key facet of the National Nanotechnology Initiative has been the development of an extensive infrastructure of interdisciplinary research and education centers, networks, and user facilities. The centers and user facilities are located at universities and federal laboratories across the country.

Centers and networks provide opportunities and support for multidisciplinary research among investigators from a variety of disciplines and research sectors, including academia, industry, and government laboratories. Such multidisciplinary research not only can lead to advances in knowledge, but also may foster relationships that further the development of basic research results into devices and other applications.

Many agencies support such centers. Examples of federal and federally supported centers include:

- The National Science Foundation has established university-based centers focused exclusively on nanotechnology, including 19 Nanoscale Science and Engineering Centers (NSECs), 13 of which continue funding in FY2013; one Engineering Research Center; one Science and Technology Center; four Materials Research Science and Engineering Centers fully dedicated to nanotechnology and 19 others with one or more interdisciplinary research group(s) focused on nanoscale science and engineering topics; two Nanoscale Science and Engineering Education Centers; and five Nanoscale Science and Engineering Networks.[60] NSF anticipates the establishment of three Nanosystems Engineering Research Centers (NERCs) in 2012, becoming fully operational in 2013.[61]

- The NIH has established more than 28 centers, including eight university-based Nanomedicine Development Centers; a Nanotechnology Characterization Laboratory, established by the National Cancer Institute (NCI), in partnership with NIST and the Food and Drug Administration; nine university-based Centers of Cancer Nanotechnology Excellence, established under the NCI's Alliance for Nanotechnology in Cancer initiative; four university-based centers, established by the National Heart, Lung, and Blood Institute under its Program of Excellence in Nanotechnology; and six university-based NCI Cancer Nanotechnology Training Centers.[62]

- The Department of Defense supports three university-based nanotechnology research centers, as well as the Institute for Nanoscience at the Naval Research Laboratory.

- The Department of Energy has established five Nanoscale Science Research Centers (NSRCs) co-located with its national labs.

- NIST has established a Center for Nanoscale Science and Technology (CNST).

---

[60] NNI website, *Centers and Networks*, http://www.nano.gov/centers-networks.

[61] NSET Subcommittee, NSTC, EOP, *The National Nanotechnology Initiative: Research and Development Leading to a Revolution in Technology and Industry, Supplement to the President's FY2013 Budget*, February 2012.

[62] NNI website, *Centers and Networks*, http://www.nano.gov/centers-networks.

- NIOSH has established a Nanotechnology Research Center to conduct research into the application of nanoparticles and nanomaterials in occupational safety and health and the implications of nanoparticles and nanomaterials for work-related injury and illness.

Many of the centers are designated as user facilities and are available to researchers not located at the center. User facilities are designed to allow outside researchers to take advantage of facilities, equipment, tools, and expertise. These shared resources provide researchers the opportunity to conduct research, characterize materials, and test products using equipment and facilities that their individual companies, universities, or organizations could not afford to acquire, support, or maintain. Conditions for user access vary by facility and agency. In general, users are not charged for pre-competitive, non-proprietary work leading to publication, and are charged on a cost-recovery basis for proprietary work. In some cases, the user facilities are located at federal government laboratories (e.g., the Department of Energy's NSRCs, and the NIST CNST); other user facilities are located at universities and supported with federal funds (e.g., NSF's university-based centers in the National Nanotechnology Infrastructure Network (NNIN)).

As mentioned earlier, the 21st Century Nanotechnology R&D Act of 2003 directed the establishment of two centers, the American Nanotechnology Preparedness Center and the Center for Nanomaterials Manufacturing. According to the NSET Subcommittee, the requirement to establish the American Nanotechnology Preparedness Center was met by NSF's establishment of the Network for Nanotechnology in Society, comprised of centers at the University of California, Santa Barbara (with the participation of Harvard University and the University of South Carolina) and the University of Arizona.[63] These centers were funded under NSF's Nanoscale Science and Engineering Center (NSEC) program and did not include participation by any other NSET Subcommittee agency.[64] The NSET Subcommittee states that the requirement for establishing the Center for Nanomaterials Manufacturing was met by NSF's establishment of a National Nanomanufacturing Network (NNN) comprised of four NSECs. The Center for Integrated Hierarchical Manufacturing at the University of Massachusetts Amherst is the main node of the NNN.[65] The NNN NSECs were established by NSF in collaboration with DOD and NIST, but exclusively with NSF funds.[66]

NNI Assessments by PCAST, NRC: Selected Issues, Findings, and Recommendations

The 21st Century Nanotechnology R&D Act of 2003 (P.L. 108-153) requires periodic external reviews of the National Nanotechnology Program (NNP) by the National Research Council (NRC) of the National Academy of Sciences (NAS),[67] and by the National Nanotechnology Advisory Panel (NNAP), directing the President to "establish or designate a National

---

[63] Private telephone communication between CRS and NSTC staff, January 31, 2008.

[64] Private e-mail communication between CRS and NSF staff, January 31, 2008.

[65] Private telephone communication between CRS and NSTC staff, January 31, 2008.

[66] Private e-mail communication between CRS and NSF staff, January 31, 2008.

[67] The NAS and NRC are part of the National Academies which also includes the National Academy of Engineering and the Institute of Medicine.

Nanotechnology Advisory Panel." Both President Obama[68] and President Bush[69] issued executive orders designating the President's Council of Advisors on Science and Technology to serve as the NNAP. References to PCAST in this section refer to the council operating in its capacity as the NNAP.

Under the act, the NNAP is to report to the President and Congress on its assessments of the NNP at least once every two years on a variety of factors, including: trends and developments in nanotechnology science and engineering; progress made in implementing the NNP; the need to revise the program; the balance among the components of the program, including funding levels for the program component areas; whether the program component areas, priorities, and technical goals developed by the NSTC are helping to maintain U.S. leadership in nanotechnology; the management, coordination, implementation, and activities of the program; and whether societal, ethical, legal, environmental, and workforce concerns are adequately being addressed.

In addition, the act directs the NNCO to enter into an arrangement with NRC to conduct a triennial review of the program. Each review is to include an evaluation of the technical accomplishments of the program, including a review of whether the program has achieved the goals under the metrics established by the NSTC; a review of the program's management and coordination across agencies and disciplines; a review of the funding levels at each agency for the program's activities and the ability of each agency to achieve the program's stated goals with that funding; an evaluation of the program's success in transferring technology to the private sector; an evaluation of whether the program has been successful in fostering interdisciplinary research and development; an evaluation of the extent to which the program has adequately considered ethical, legal, environmental, and other appropriate societal concerns; recommendations for new or revised program goals; recommendations for new research areas, partnerships, coordination and management mechanisms, or programs to be established to achieve the program's stated goals; recommendations on policy, program, and budget changes with respect to nanotechnology research and development activities; recommendations for improved metrics to evaluate the success of the program in accomplishing its stated goals; a review of the performance of the NNCO and its efforts to promote access to and early application of the technologies, innovations, and expertise derived from program activities to agency missions and systems across the federal government and to U.S. industry; an analysis of the relative position of the United States compared to other nations with respect to nanotechnology R&D, including the identification of any critical research areas where the United States should be the world leader to best achieve the goals of the program; and an analysis of the current impact of nanotechnology on the U.S. economy and recommendations for increasing its future impact.

The PCAST has produced four assessments of the NNI; the NRC has produced one:

- *The National Nanotechnology Initiative at Five Years: Assessment and Recommendations of the National Nanotechnology Advisory Panel*, NNAP/PCAST, May 2005 (herein referred to as "*First Assessment*").

- *A Matter of Size: Triennial Review of the National Nanotechnology Initiative*, NAS/NRC, 2006.

---

[68] Executive Order 13539.

[69] Executive Order 13349.

- *The National Nanotechnology Initiative: Second Assessment and Recommendations of the National Nanotechnology Advisory Panel,* NNAP/PCAST, April 2008 (herein referred to as *"Second Assessment"*).

- *Report to the President and Congress on the Third Assessment of the National Nanotechnology Initiative,* NNAP/PCAST, March 2010, (herein referred to as *"Third Assessment"*).

- *Report to the President and Congress on the Fourth Assessment of the National Nanotechnology Initiative,* NNAP/PCAST, April 2012 (herein referred to as *"Fourth Assessment"*).

The NRC is currently engaged in its second assessment. Selected findings and recommendations of these assessments, as well as NSET's responses to PCAST's *Third Assessment* recommendations, are grouped by broad issue area and discussed below.

NNI Program Management

In its *Third Assessment*, PCAST praised the NNI for having "distinguished itself during its first decade as a successful cooperative venture," and further described the initiative as well-organized and well-managed.[70] Nevertheless in both its *Third Assessment* and *Fourth Assessment*, PCAST offered a number of recommendations for improving NNI program management. In its *Third Assessment*, PCAST stated that "the NNCO should broaden its impact and efficacy and improve its ability to coordinate and develop NNI programs and policies." In this regard, PCAST recommended that OSTP undertake the following actions:

- Require each agency in the NNI to have senior representatives with decision-making authority participate in coordination activities of the NNI.

- Strengthen the NNCO to enhance its ability to act as the coordinating entity for the NNI.

- Dedicate 0.3% of NNI funding to the NNCO to ensure the appropriate staffing and budget to effectively develop, monitor and assess NNI programs.[71]

In its 2013 budget supplement, NSET Subcommittee characterized these program management recommendations as actions "unlikely or not needed." With respect to requiring the participation of senior agency representatives with decision-making authority, the NSET Subcommittee replied that member agencies make their own decisions with regard to representation. The NSET Subcommittee also responded to PCAST's recommendation that the NNCO's coordination function be strengthened and better funded. First, the NSET Subcommittee reasserted that it is the NSET Subcommittee, not the NNCO, that serves as the coordinating entity for the NNI, and that the NNCO serves to provide administrative support to the NSET Subcommittee. Second, the NSET Subcommittee stated that it did not support providing the NNCO a fixed percentage of the overall NNI investment, explaining that NNCO activities are based on programmatic needs as proposed each year by the NNCO and vetted by the NSET Subcommittee's Budget Steering Group.

---

[70] PCAST, EOP, *Third Assessment,* March 12, 2010.

[71] Ibid. Two other PCAST program management recommendations related to metrics and appointment of individuals to the NNCO to lead interagency EHS and standards development efforts are discussed elsewhere in this section.

In its *Fourth Assessment*, PCAST lauded the NNI's "significant progress" in addressing several issues identified in the *Third Assessment*, but expressed concern about the lack of progress in others, notably "significant and persistent hurdles to an optimal structure and management of the initiative." In particular, PCAST noted that despite the publication in 2011 of the National Nanotechnology Initiative Strategic Plan,[72]

> individual agency contributions lack the cohesion of an overarching framework, and there is no clear connection between the goals and objectives of the NNI strategic plan with those of individual agencies.[73]

While recognizing that some agencies do not have dedicated nanotechnology programs, but rather decentralized nanotechnology activities across multiple organizational units, PCAST recommended the development of implementation plans by each agency—in consultation with external stakeholders—that discuss the alignment of agency activities with the objectives of the NNI strategic plan.

PCAST renewed its recommendations that each NSET agency have senior representatives with decision-making authority participate in coordination activities of the NNI, and that NSET dedicate 0.3% of NNI funding to the NNCO to ensure the appropriate staffing and budget to effectively develop, monitor, and assess NNI programs (raising NNCO funding from approximately $3 million to $5 million).[74]

To expand and strengthen the role of the NNCO in the NNI, PCAST also recommended appointing the NNCO director as co-chair of the NSET Subcommittee and allowing non-federal experts to serve as NNCO director.[75]

PCAST further recommended creating a standing PCAST Nanotechnology Steering Committee of experts from industry, academia, and civil society to provide more frequent and in-depth guidance.[76] This recommendation raises the question of whether the role envisioned for the NNAP, as authorized by the 21st Century Nanotechnology Research and Development Act, is appropriately being served by PCAST or whether it might be better served by the establishment of a separate, dedicated entity. In its assessment of the NNI, the NRC recommended that

> the federal government [should] establish an independent advisory panel with specific operational expertise in nanoscale science and engineering; management of research centers, facilities, and partnerships; and interdisciplinary collaboration to facilitate cutting-edge research on and effective and responsible development of nanotechnology.[77]

Coming after President Bush's designation of PCAST to serve as the legislatively mandated NNAP, this recommendation may suggest the need for a separate, NNI-only focused entity to serve as the NNAP. Critics of the use of PCAST to serve as the NNAP maintain that the scope and depth of expertise needed to provide effective guidance on the NNI requires an independent

---

[72] NSTC Subcommittee, NSTC, EOP, *National Nanotechnology Initiative Strategic Plan*, February 2011, http://nano.gov/sites/default/files/pub_resource/2011_strategic_plan.pdf.

[73] PCAST, EOP, *Fourth Assessment*, April 2012.

[74] Ibid.

[75] Ibid

[76] Ibid.

[77] NRC, *A Matter of Size: Triennial Review of the National Nanotechnology Initiative*, 2006.

panel of people with nanotechnology—specific and interdisciplinary expertise and an undivided focus. Supporters of using PCAST for this function assert that a single advisory panel provides an integrated perspective, reduces unnecessary cost and management burdens, and allows for expertise to be added to the panel or accessed through non-member technical advisory groups.

Funding

A key question for federal policymakers is how much funding should be provided for nanotechnology R&D, and how should this funding be directed among program component areas and cross-cutting activities. NNI funding has grown from $464 million in FY2001 to an estimated $1.697 billion in FY2012, however regular appropriations for the NNI have fallen $216 million (-11.3%) from their peak in FY2010 ($1.913 billion).

In its *Third Assessment*, PCAST recommended that the federal government increase funding for the NNI to ensure that the United States retains its

> leadership role in the development and commercialization of nanotechnology in the face of mounting competition from countries that have responded to the example set by the NNI.[78]

PCAST cautioned that in undertaking these new investments the NNI should maintain or expand the level of funding devoted to basic nanotechnology research.

Similarly, in its assessment of the NNI, the NRC recommended

> the federal government [should] sustain investments in a manner that balances the pursuit of shorter-term goals with support for longer-term R&D and that ensures a robust supporting infrastructure, broadly defined. Supporting long-term research effectively will require making new funds available that do not come at the expense of much-needed ongoing investment in U.S. physical sciences and engineering research.[79]

U.S. Technological and Industrial Leadership

Given the economic, societal, and national security potential of nanotechnology, Congress maintains ongoing interest in the competitive position of the United States in this emerging field. In its *Fourth Assessment,* PCAST asserted that the NNI "remains a successful cooperative venture" that has had a "catalytic and substantial impact" on the growth of the U.S. nanotechnology industry, and concluded that

> the United States is today, by a wide range of measures, the global leader in this exciting and economically promising field of research and technological development.[80]

---

[78] PCAST, EOP, *Report to the President and Congress on the Third Assessment of the National Nanotechnology Initiative*, March 12, 2012, http://www.whitehouse.gov/sites/default/files/microsites/ostp/pcast-nni-report.pdf.

[79] NRC, *A Matter of Size: Triennial Review of the National Nanotechnology Initiative*, 2006, http://books nap.edu/catalog.php?record_id=11752.

[80] PCAST, EOP, *Report to the President and Congress on the Fourth Assessment of the National Nanotechnology Initiative*, April 2012, http://www.whitehouse.gov/sites/default/files/microsites/ostp/PCAST_2012_Nanotechnology_FINAL.pdf.

PCAST attributed the U.S. leadership position in nanotechnology, in large part, to the NNI.

The assertion of U.S. leadership in nanotechnology echoes PCAST's findings in its *Third Assessment*. That assessment, however, noted that U.S. leadership was threatened by several aggressive competitors:

> The United States is clearly the world's leader in nanotechnology R&D and commercialization based on research funding, total number of papers in the most significant scientific publications, patents filed and granted, private sector funding for new and existing companies developing nanotechnologies, and sales of nanotechnology-based products. However, foreign competitors, particularly China, South Korea, Germany, and Japan, are making gains on many of these same metrics. China in particular has significantly increased its share of nanotechnology research publications and patents and now supports nanotechnology as a larger fraction of its total scientific research compared to the United States....Though still the leader in nanotechnology R&D and commercialization, the United States is losing ground to foreign competitors, particularly China, South Korea, Germany, and Japan, on a number of key metrics of research output and commercial activity. [81]

PCAST's *Fourth Assessment* noted that "in addition to China, South Korea, and other early movers, the Russian Nanotech Corporation (RUSNANO) is now also rising as a major player, second only to the United States in its nanotechnology R&D spending."[82] PCAST cited data from Lux Research indicating that RUSNANO had increased its funding by nearly 40% to $1.05 billion, and planned to increase funding to nearly $1.5 billion by 2015.

PCAST's assessment of the U.S. leadership position is founded not on sales, growth, or market share of commercial products—common measures of global competitiveness for established products—but rather on metrics that may serve as early indicators of potential innovation, such as the U.S. share of scientific publications and patents. The use of such metrics may not be universally accepted as predictive of leadership position. Technological leadership—or even leadership in innovation—does not ensure that the economic benefits from such leadership will accrue to the United States. Companies may choose to manufacture products or conduct other value-added activities outside the United States. If the assessment of national competitiveness is expanded to include value-added activities and jobs generated or retained within the United States, then the metrics for assessing leadership might change.

Technology Commercialization

Technology commercialization involves the movement of scientific and technological insights into products and services. It is the process by which knowledge created by investments in R&D is translated into economic benefits (e.g., strengthening existing firms and establishing new ones, creating jobs, producing new products, reducing the cost of existing products and improving their performance, delivering returns to investors) and societal benefits (e.g., offering new and improved sources of renewable energy, reducing pollution and remediating environmental damage, improving human heath).

---

[81] PCAST, EOP, *Third Assessment*, March 12, 2010.

[82] PCAST, EOP, *Fourth Assessment*, April 2012.

In its *Third Assessment*, PCAST recommended that, to maintain the U.S. leadership position in nanotechnology,

> the NNI increase its emphasis on nanomanufacturing and commercial deployment of nanotechnology-enabled products, and that the agencies within the NNI must interact and cooperate more with one another to ease the translation of scientific discovery into commercial activity.[83]

In this regard, PCAST made a number of recommendations. Among them:

- double federal funding for nanomanufacturing over five years;

- launch at least five government-industry-university partnerships modeled after the Nanoelectronics Research Initiative;

- for each Nanotechnology Signature Initiative (NSI), develop milestones, promote strong education components, and create public-private partnerships to leverage the outcomes;

- fund at least five NSIs over the next two to three years—including ones in priority areas such as homeland security, national defense, and human health—at annual levels of $20 million to $40 million each;

- tap the Department of Commerce and the Small Business Administration for advice on how the NNI can best ensure its programs create new jobs in the United States, including mechanisms for coordinating with state efforts; and

- for DOE, DOD, NIST, NIH, NCI, and FDA to clarify the development pathway and to make sustained investments to accelerate technology transfer to the marketplace.[84]

In its 2013 budget supplement, NSET responded to PCAST's commercialization-focused recommendations stating that "key NNI agencies" were on track to double their nanomanufacturing investments over five years, while maintaining their investments in fundamental nanotechnology research. NSET also said that interagency task forces for the three initial NSIs were developing coordination plans with milestones, that there was a likelihood that two additional NSIs would be added within the next year, and that its NILI working group was working in conjunction with the NNCO's Industry and State liaison to produce a work plan that includes job creation and outreach to states and industry. With respect to clarifying the development pathway, NSET asserts that it is employing well-established mechanisms, such as the Small Business Innovation Research (SBIR) program and the Small Business Technology Transfer (STTR) program, and are developing new programs to accelerate technology transfer and to clarify regulatory pathways.[85]

In its *Fourth Assessment*, PCAST continued its emphasis on an enhanced focus on commercialization efforts, and lauded the NNI's response to commercialization-related recommendations the council made in its previous assessment, specifically:

---

[83] PCAST, EOP, *Third Assessment*, March 12, 2010.

[84] Ibid.

[85] NSET Subcommittee, NSTC, EOP, *The National Nanotechnology Initiative: Supplement to the President's 2013 Budget*, February 2012.

- development of an NNCO Industry and State Liaison position and expanded efforts by the NNCO in supporting nanotechnology commercialization;

- the NILI working group's development of an agenda focused on job creation and state outreach, as well as mechanisms to incorporate industrial input in NNI planning;

- DOE programs that include industrial partners to overcome technological barriers to nanotechnology commercialization;

- NIST's plans to start the Advanced Manufacturing Technology Consortia in FY2013 to speed development and commercialization of new products and services, including nanotechnology; and

- NIH's creation of the National Center for Advancing Translational Sciences to accelerate translation of promising technologies and clinical studies.[86]

Nevertheless, the PCAST co-chairs also noted in their transmittal letter to President Obama that additional work in a number of areas was still required to facilitate commercialization and U.S. leadership in nanotechnology:

> ... additional efforts are needed in four areas: strategic planning, program management, metrics for assessing impact, and increasing support for research on environmental, health, and safety issues associated with nanotechnology. Continued lack of attention to these concerns will make it harder for the U.S. to maintain its leadership role in the commercialization of nanotechnology.[87]

Standards

Standards are likely to play a critical role in many aspects of nanotechnology R&D and commercialization, among them the development of research tools, conduct of research, reproducibility of experimental results, development and enforcement of regulations, materials characterization, nanomanufacturing, and product testing and evaluation.

In its *Second Assessment*, PCAST found that

> Progress across the breadth of NNI-supported R&D critically depends upon the development and implementation of standards for nanomaterial identification, characterization, and risk assessment.[88]

In its *Third Assessment*, PCAST re-asserted the importance of standards stating that "the establishment of standards is essential to growth of most new technologies, and nanotechnology is no exception,"[89] and recommending that the NNCO serve as the coordinating agency for

---

[86] PCAST, EOP, *Fourth Assessment,* April 2012.

[87] Ibid.

[88] PCAST, EOP, *Second Assessment,* April 2008.

[89] PCAST, EOP, *Third Assessment,* March 12, 2010.

collaborating with stakeholders on enabling programs such as metrology; standards including size, shape and composition of nanomaterials, and databases of physical and chemical properties of nanomaterials; and manufacturing safety. [90]

In support of this role, PCAST also recommended that an individual be appointed to the NNCO to lead interagency coordination of standards development efforts.[91] In its 2013 budget supplement, NSET responded that it had appointed then-NNCO Director Clayton Teague to serve in this capacity.[92]

Economic Impact Metrics and Data Collection

In its *Third Assessment,* PCAST highlighted the need for economic impact metrics and data collection, recommending the development of NNI economic impact metrics; making economic impact an explicit metric in the second decade of the NNI; and lodging responsibility with a statistical agency (such as the DOC's Bureau of Economic Analysis) to estimate job creation and the value of nanotechnology products and products incorporating nanotechnology components, rather than relying on funding agencies for such estimates.[93]

In its 2013 budget supplement, NSET responded that the NNCO had engaged in discussions with the Department of Commerce about economic metrics, provided support for a symposium on the economic value of nanotechnology, and requested the NRC identify metrics to assess the success of nanotechnology as part of the council's triennial review of the NNI.[94]

In its 2006 assessment, the NRC also recommended a focus on development of metrics and a greater role for the DOC in economic data collection and analysis:

> [The NRC recommends] the NSET Subcommittee carry out or commission a study on the feasibility of developing metrics to quantify the return to the U.S. economy from the federal investment in nanotechnology R&D. The study should draw on the Department of Commerce's expertise in economic analysis and its existing ability to poll U.S. industry. Among the activities for which metrics should be developed and relevant data collected are technology transfer and commercial development of nanotechnology.[95]

Few efforts have been made within the federal government to understand the economic impacts of the nation's investments in the NNI. Identification and tracking of data that could serve as an indicator of success in commercializing nanotechnology research or the effects on U.S. job creation or retention has not been formalized. To the extent that federal assessments of the economic contribution of and/or potential for nanotechnology products have occurred, they have not been performed with analytical rigor. Although the Commerce Department retains its economic analysis expertise, resident primarily in the Economics and Statistics Administration's

---

[90] Ibid.

[91] Ibid.

[92] NSET Subcommittee, NSTC, EOP, *The National Nanotechnology Initiative: Supplement to the President's 2013 Budget,* February 2012.

[93] PCAST, EOP, *Third Assessment,* March 12, 2010.

[94] NSET Subcommittee, NSTC, EOP, *The National Nanotechnology Initiative: Supplement to the President's 2013 Budget,* February 2012.

[95] NRC, *A Matter of Size: Triennial Review of the National Nanotechnology Initiative,* 2006.

Bureau of Economic Analysis, the department's Technology Administration, which led Commerce's NNI activities and had government-wide responsibilities for technology transfer activities, was eliminated in August 2007.[96] Prior to its elimination, the Technology Administration contracted for two studies that could contribute to addressing this NRC recommendation: an analysis of barriers to nanotechnology commercialization performed by the University of Illinois at Springfield, and an analysis of innovation metrics conducted by the Alliance for Science and Technology Research in America (ASTRA). These reports are publicly available at Commerce Department websites.[97]

Nanotechnology Workforce Education and Training

With nanotechnology advocates promising the creation of many new jobs—some have estimated the number to be in the millions—as a result of global nanotechnology investments, some have asserted that the country must prepare students for nanotechnology research, engineering, and production jobs.[98] Assessing which industries are likely to create such jobs, which skills will be needed, and in what timeframe are key challenges. If workers with nanotechnology-specific skills are needed and no workers are available domestically (e.g., U.S. citizens, resident aliens, or those in the United States on work visas), potential employers may opt to establish or move operations outside the United States to tap workers with those skills abroad. Conversely, if students are trained for jobs that do not emerge or do not emerge in the same timeframe as students are entering the job market, this investment may be lost. In addition, potential students may be discouraged from pursuing future nanotechnology-related studies.

In its *Third Assessment,* PCAST found that the United States remained unchallenged in educating nanotechnology researchers. According to PCAST, NSF supports the training and education of about 10,000 students and teachers in nanoscale science and engineering, funds the development of new curricula for nanotechnology education, and is expanding the outreach efforts of the National Center for Nanotechnology Applications and Career Knowledge. However, PCAST also noted that a large number of foreign students return to their home countries after completing their education.

> The United States still trains the majority of Ph.D. students in nanoscience and nanotechnology, and though many of these students wish to remain in the United States after completing their degree programs, the data show that over one-third of these students return to their home countries and contribute to the development of nanotechnology R&D programs throughout the world. [99]

While acknowledging that the United States may gain some benefits from training nanotechnology researchers that return to their home countries, PCAST recommended the federal

---

[96] The Technology Administration was eliminated in the America COMPETES Act (P.L. 110-69).

[97] College of Business Management, University of Illinois at Springfield, *Barriers to Nanotechnology Commercialization*, September 2007, http://www.osec.doc.gov/Report-Barriers%20to%20Nanotechnology%20Commercialization.pdf; and ASTRA, *Innovation Vital Signs Project*, July 2007. http://www.ntis.gov/ta_reports/Report-InnovationVitalSigns.pdf.

[98] Phillip J. Bond, Under Secretary for Technology, U.S. Department of Commerce, remarks, "Nanotechnology: Economic Opportunities, Societal and Ethical Challenges," NanoCommerce 2003, December 9, 2003. http://www.technology.gov/Speeches/PJB_031209 htm *Sizing Nanotechnology's Value Chain*, Lux Research, October 2004.

[99] PCAST, EOP, *Third Assessment*, March 12, 2010.

government undertake efforts to retain scientific and engineering talent trained in the United States

> by developing a program to provide U.S. Permanent Resident Cards for foreign individuals who receive an advanced degree in science or engineering at an accredited institution in the United States and for whom proof of permanent employment in that scientific or engineering discipline exists.[100]

PCAST also recommended that NNI agencies continue making investments in innovative and effective education, and that the NNCO should consider commissioning a comprehensive evaluation of the outcomes of the overall investment in NNI education.[101]

In its 2013 budget supplement, NSET responded to PCAST's *Third Assessment* workforce education and training recommendations stating that NNI agencies were contributing to the development of nanotechnology curricula for students in grade school through postdoctoral training, and that NSF is considering supporting an external study to evaluate the NNI's investment in education. In addition, while acknowledging the need to undertake efforts to retain U.S. educated foreign scientific and engineering talent in the United States, NSET stated that it did not endorse specific approaches at this time. [102]

Close coordination among the Departments of Commerce, Education, and Labor might help to align federal education and training efforts better with the labor market for nanotechnology workers. In its 2006 assessment, the NRC recommended the NSET Subcommittee "create a working group on education and the workforce that engages the Department of Education and Department of Labor as active participants."[103] An NSET Subcommittee working group on education and the workforce has not been established.

Environmental, Health, and Safety-Related Issues

Environmental, health, and safety issues related to nanotechnology research, development, use, and disposal continue to be a focus of NNI assessments. Some analysts have described nanotechnology as a two-edged sword. On the one hand, some are concerned that nanoscale particles may enter and accumulate in vital organs, such as the lungs and brains, potentially causing harm or death to humans and animals, and that the diffusion of nanoscale particles in the environment might harm ecosystems. On the other hand, some analysts believe that nanotechnology has the potential to deliver important EHS benefits such as reducing energy consumption, pollution, and greenhouse gas emissions; remediating environmental damage; curing, managing, or preventing diseases; and offering new safety-enhancing materials that are stronger, self-repairing, and able to adapt to provide protection.

Stakeholders generally agree that concerns about potential detrimental effects of nanoscale materials and devices—both real and perceived—must be addressed to protect and improve human health, safety, and the environment; create public faith and confidence in the safety of

---

[100] Ibid.

[101] Ibid.

[102] NSET Subcommittee, NSTC, EOP, *The National Nanotechnology Initiative: Supplement to the President's 2013 Budget*, February 2012.

[103] NRC, *A Matter of Size: Triennial Review of the National Nanotechnology Initiative*, 2006.

nanotechnology products; enable accurate and efficient risk assessment, risk management, and cost-benefit trade-offs; reduce EHS and related regulatory uncertainties that may impede investment;[104] foster innovation; and ensure that society can enjoy the widespread economic and societal benefits that nanotechnology may offer.

PCAST's *Third Assessment* found that:

> In the absence of more detailed scientific evidence—and effective assessment and communication of the evidence that does exist—the distinction between plausible and implausible risks remains unclear. The resulting uncertainty threatens to undermine confidence and trust among investors, businesses, and consumers, and could jeopardize the success of nanotechnology. This is not a hypothetical threat. Consumer and advocacy groups already have raised concerns over the use of engineered nanomaterials in products as diverse as clothing, fuel additives, and sunscreens. Businesses have been hampered by regulatory uncertainty. A number of industries have shied away from nanotechnology for fear of consumer rejection in the face of speculative concerns.[105]

In its 2006 assessment, the National Research Council recommended expansion of funding for EHS nanotechnology research. The NRC specifically noted the need for assessing the effects of engineered nanomaterials on public health and the environment and recommended the development of effective methods to estimate the exposure of humans, wildlife, and other ecological receptors to source material; assess effects on human health and ecosystems of both occupational and environmental exposure; and characterize, assess, and manage the risks associated with exposure.[106] While the NRC asserted the need for additional EHS research, it did not quantify how much more was needed.

In its *Third Assessment,* PCAST credited the NNI with increasing funding for EHS-focused research. Funding for EHS-related nanotechnology research across all agencies grew from $37.7 million in FY2006 to $90.2 million in FY2010. In FY2012, EHS funding is estimated to total $102.7 million.

PCAST also indentified six potential EHS hurdles facing the NNI: leadership and accountability in identifying and addressing cross-cutting issues and ensuring a sound, risk-based approach to R&D and applications; more active engagement with stakeholders; connecting research to decision-making; framing the EHS risk issue; development of a clear multi-stakeholder research strategy; and targeted funding for EHS research.[107] With respect to these hurdles, PCAST made four EHS-related recommendations:

> Risk Identification: The NSET Subcommittee's NEHI working group should develop clear principles to support the identification of plausible risks associated with the products of nanotechnology.

> Strategic Planning: The NSET Subcommittee's NEHI working group should further development and implement a cross-agency strategic plan that links EHS research activities with knowledge gaps and decision-making needs within government and industry.

---

[104] For more information on regulatory issues associated with nanotechnology, see CRS Report RL34332, *Engineered Nanoscale Materials and Derivative Products: Regulatory Challenges*, by Linda-Jo Schierow.

[105] PCAST, EOP, *Third Assessment,* March 12, 2010.

[106] NRC, *A Matter of Size: Triennial Review of the National Nanotechnology Initiative*, 2006.

[107] PCAST, EOP, *Third Assessment,* March 12, 2010.

Organizational Changes: The NSET Subcommittee and OSTP should foster administrative changes and communications mechanisms ... [by assigning] an individual to NNCO to oversee interagency efforts than address nanotechnology EHS; [expanding] the NEHI charter to enable the group to address cross-agency nanotechnology-related policy issues more broadly; and [exploring] mechanisms that enable the NEHI working group to more effectively receive input and advice from nongovernment experts in the field of emergent risks.

Information Resources: The NSET Subcommittee's NEHI working group should develop information resources on cross-cutting nanotechnology EHS issues that are relevant to businesses, health and safety professionals, researchers, and consumers.[108]

In its 2013 budget supplement, NSET responded to PCAST's *Third Assessment* EHS recommendations by noting its:

- release of *NNI Environmental, Health, and Safety (EHS) Research Strategy*,[109] which it asserted provides clear principles to guide identification of plausible risks;

- designation of a individual within the NNCO to lead interagency coordination efforts in EHS research;

- NEHI working group's establishment of a framework for taking comments from stakeholders and providing information on crosscutting EHS issues; and

- NEHI working group's coordination with the administration's recently established Emerging Technology Interagency Policy Committee.[110]

PCAST's *Fourth Assessment* lauded the "significant progress" made by the NNI to address potential EHS risks of nanotechnology, noting the rapid growth rate for EHS-focused research funding compared to overall NNI funding; the implementation of PCAST's earlier recommendation to identify a central coordinator for EHS-research within the NNCO; and for development and release of an EHS research strategy, articulating an approach that incorporates the "evolving research needs and the strategic research plans of three relevant agencies."[111] PCAST noted that the NNI strategy aligned with the findings of a January 2012 NRC report, *A Research Strategy for Environmental, Health, and Safety Aspects of Engineered Nanomaterials*,[112] specifically with respect to

the importance of a life cycle approach to assessing risks, the need for more research on human and environmental exposure to nanomaterials, better tools for measuring and tracking

---

[108] Ibid.

[109] NSET Subcommittee, NSTC, EOP, *NNI Environmental, Health, and Safety (EHS) Research Strategy*, 2011, http://www nano.gov/sites/default/files/pub_resource/nni_2011_ehs_research_strategy.pdf.

[110] NSET Subcommittee, NSTC, EOP, *The National Nanotechnology Initiative: Supplement to the President's 2013 Budget*, February 2012.

[111] PCAST, EOP, *Fourth Assessment*, April 2012.

[112] NRC, Committee to Develop a Research Strategy for Environmental, Health, and Safety Aspects of Engineered Nanomaterials, *A Research Strategy for Environmental, Health, and Safety (EHS) Aspects of Engineered Nanomaterials (ENMs)*, 2012, http://www.nap.edu/catalog.php?record_id=13347.

nanomaterials, and the need for cross-cutting informatics infrastructure for nanotechnology-related EHS research.[113]

However, PCAST expressed concerns about "a lack of integration between nanotechnology-related EHS research funded through the NNI and the kind of information policymakers need to effectively manage potential risks from nanomaterials."[114] To address this concern, PCAST recommended that OSTP and the NSET Subcommittee expand the charter of the NEHI working group to enable the group to address cross-agency nanotechnology-related policy issues more broadly, and that the NNI should ensure close integration of its efforts with those of the Emerging Technologies Interagency Policy Coordination Committee (ETIPC).[115] PCAST also recommended additional funding for cross-cutting areas of EHS that promote knowledge transfer such as informatics, partnerships, and instrumentation development.[116]

Some advocates for increased focus on nanotechnology-related EHS issues have proposed the establishment of a separate agency or office devoted to nanotechnology EHS research, and/or to set aside a particular percentage of NNI funding for EHS research In its *Second Assessment*, PCAST found these proposals to be "misguided" and potentially counterproductive as such approaches may direct resources away from research "on beneficial applications and on risk." The panel also concluded that nanotechnology does not raise ethical concerns unique from those accompanying other technological advances.[117]

Societal Implications

The term "societal implications" in the context of the NNI refers to the effects, broadly speaking, that advances in nanotechnology research and application may have on individuals, groups, and society as a whole. With nanotechnology holding potential breakthroughs in areas such as materials and manufacturing, medicine and healthcare, environment and energy, biotechnology and agriculture, electronics and information technology, and national security,[118] the societal implications—including ethical, economic, and legal implications—may be both deep and widespread.

Understanding the potential societal implications of nanotechnology is considered important to federal efforts to maximize nanotechnology's potential positive effects and minimize its potential negative effects. Beginning with its first review of the NNI, PCAST stressed the importance of research aimed at understanding the societal implications of nanotechnology and recommended

---

[113] PCAST, EOP, *Fourth Assessment*, April 2012.

[114] Ibid.

[115] The Emerging Technologies Interagency Policy Coordination Committee was created jointly in 2010 by OSTP, OMB's Office of Information and Regulatory Affairs, and the Office of the United States Trade Representative. The ETIPC members include assistant secretary-level representatives from about 20 Federal agencies.

[116] PCAST, EOP, *Fourth Assessment*, April 2012.

[117] PCAST, EOP, *The National Nanotechnology Initiative: Second Assessment and Recommendations of the National Nanotechnology Advisory Panel*, May 2005, http://www.whitehouse.gov/sites/default/files/microsites/ostp/PCAST-NNAP-NNI-Assessment-2008.pdf.

[118] NSF, *Societal Implications of Nanoscience and Nanotechnology*, March 2001, http://www.wtec.org/loyola/nano/NSET.Societal.Implications/nanosi.pdf.

the NNI actively work to inform the public about nanotechnology and to confront societal issues in an open, straightforward, and science-based manner.[119]

PCAST has continued to stress the importance of a strong NNI program in societal implications, recommending in its *Fourth Assessment:*

> The NSET Subcommittee should develop a clear expectation and strategy for programs in the societal dimensions of nanotechnology. An effective program in societal implications would have well-defined areas of focus, clearly articulated outcomes as well as plans for assessing and evaluating those outcomes, and partnerships that leverage the value of its activities. Ultimately, the inclusion of such programs in the NNI has the goal of streamlining nanotechnology innovation and its positive impact on society, and the creation of new jobs, opportunities and a robust economy.[120]

Some critics of the NNI hold deep reservations about the ethical, societal, economic, and legal implications of nanotechnology. Some of these concerns are common to many technologies, such as the allocation of risk and benefit during manufacturing. For example, a neighborhood located near a production facility may bear risks associated with exposure to the byproducts (or products) of nanoscale manufacturing, while gaining few of the benefits. Concerns about possible adverse effects of nanoscale particles on human health and the environment resulting from their small particle size and unique characteristics may result in increased attention to such costs and benefits with respect to nanoscale material production. Currently, nanotechnology EHS risks are not well understood and may be acute, pose no more risk than other manufacturing processes, or perhaps even less.

Privacy rights are another issue associated with the products of nanotechnology. Nanotechnology may enable the production of highly sensitive, inexpensive sensors that could be deployed ubiquitously in commercial and public settings. While these sensors may allow beneficial applications, such as check-out-free purchases from stores or the monitoring of the environment for toxic substances, critics argue that they could also impinge on the privacy rights of individuals if, for example, the sensors could detect chemicals related to the use of tobacco, alcohol, or illegal substances without the permission of the individual. Such information might be later applied in law enforcement, life insurance, health insurance, or employment decisions.[121] Others express concern that the economically disadvantaged and less educated—both individuals and nations— might be unable or less able to take part in the benefits that nanotechnology products could offer.[122]

---

[119] PCAST, EOP, The National Nanotechnology Initiative at Five Years: Assessment and Recommendations of the National Nanotechnology Advisory Panel, May 2005, http://www.nano.gov/html/res/ FINAL_PCAST_NANO_REPORT.pdf.

[120] PCAST, EOP, *Fourth Assessment*, April 2012.

[121] Moore, Fiona M., "Implications of Nanotechnology Applications: Using Genetics as a Lesson," *Health Law Review*, Vol. 10, No. 3, 2002. http://www.law.ualberta.ca/centres/hli/pdfs/hlr/v10_3/10.3moorefrm.pdf.

[122] Smith, Richard H., "Social, Ethical, and Legal Implications of Nanotechnology," *Societal Implications of Nanoscience and Nanotechnology* (The Netherlands:Kluwer Academic Publishers, 2001).

Selected NNI Reports

The NNI's coordinating body, the NSTC's NSET Subcommittee, produces a variety of reports that serve to inform Congress and other key stakeholders of the initiatives' current activities, investments, and priorities. This section presents summaries of some of these reports.

The National Nanotechnology Initiative: Research and Development Leading to a Revolution in Technology and Industry, Supplement to the President's FY2013 Budget[123]

Each year the NSET Subcommittee publishes a supplement to the President's annual budget request. This report meets the annual reporting requirement of the 21st Century Nanotechnology Research and Development Act (P.L. 108-153, 15 USC §7501) as well as DOD reporting requirements under 10 USC §2358. The FY2013 NNI budget supplement provides a summary of NNI activities in FY2011 and FY2012. It also provides a detailed view of NNI funding in the President's FY2013 budget request, including a breakout of FY2011, FY2012, and proposed FY2013 funding for each program component area. The report describes proposed changes in agency R&D budgets, as well as in the balance of investments by PCA. Of particular note:

- President Obama has proposed an overall NNI budget for FY2013 of $1.767 billion, a $70.1 million (4.1%) increase from the FY2012 funding level. However, the requested level is $145.8 million (-7.6%) less than the NNI's peak funding of $1.913 billion in FY2010.

- Funding for EHS R&D in FY2013 would rise to $105.4 million, a $2.7 million (2.6%) increase over FY2012.

- Funding for nanomanufacturing R&D in FY2013 would rise to $88.9 million, a $15.4 million (21.0%) increase over FY2012.

- Funding for nanomaterials R&D in FY2013 would rise to $368.4 million, a $57.7 million (18.6%) increase over FY2012.

- Funding for fundamental phenomena and processes research in FY2013 would fall to $498.2 million, a decrease of $11.9 million (-2.3%) from FY2012.

Environmental, Health, and Safety Research Strategy[124]

The 2011 *Environmental, Health, and Safety Research Strategy* revises and replaces the 2008 strategy. The NSTC states that the strategy is "grounded in the principles of risk assessment and product life cycle analyses."[125] Risk assessment provides a process for understanding the magnitude of the potential exposure to humans and the environment and the magnitude of the

[123] NSET Subcommittee, NSTC, EOP, The National Nanotechnology Initiative: Research and Development Leading to a Revolution in Technology and Industry, Supplement to the President's FY2013 Budget, February 2012, http://nano.gov/sites/default/files/pub_resource/nni_2013_budget_supplement.pdf.

[124] NSET Subcommittee, NSTC, EOP, *Environmental, Health, and Safety Research Strategy*, October 2011, http://nano.gov/sites/default/files/pub_resource/nni_2011_ehs_research_strategy.pdf.

[125] Ibid.

potential hazard or effects presented by a nanomaterial, allowing risk-benefit comparisons among nanomaterials, between a nanomaterial and non-nanomaterials, or for a single nanomaterial. Performing such risk assessments at different stages of the life cycle of products (e.g., development, manufacture, commercialization, disposal, end-of-life), allowed for the identification of critical risk assessment data needs. These needs were then translated into nanotechnology-related EHS research needs, and organized into six categories: (1) Nanomaterial Measurement Infrastructure, (2) Human Exposure Assessment, (3) Human Health, (4) Environment, (5) Risk Assessment and Risk Management Methods, and (6) Informatics and Modeling. For each category, the strategy provided an overview, specified a goal, and identified the current research needs. The strategy also included a chapter on implementation, including targeting and accelerating EHS research, linking EHS research needs and goals to the NNI Strategic Plan, and disseminating EHS research needs and knowledge.

The National Nanotechnology Initiative Strategic Plan (2011)[126]

The 21st Century Nanotechnology R&D Act of 2003 (P.L. 108-153) requires the NSTC to develop an NNI strategic plan every three years. This plan is to guide the program's activities to meet the goals, priorities, and anticipated outcomes of the participating agencies. In addition, the act requires the triennial strategic plan to address how the program intends to move results out of the laboratory and into application for the benefit of society, its plan for long-term funding for interdisciplinary R&D, and the allocation of funding for interagency projects.

The 2011 *National Nanotechnology Strategic Plan* maintains the overall vision, four goals, and eight program component areas of the 2007 NNI Strategic Plan.[127] The 2011 plan highlights the roles and interests of NNI participating agencies, establishes specific objectives for each of the four NNI goals, and includes a chapter on future directions for the NNI. In particular, the report highlights the Administration's NNI "signature initiatives," multi-agency efforts

> intended to enable the rapid advancement of science and technology in the service of national economic, security, and environmental goals by focusing resources on critical challenges and R&D gaps.

These initiatives include Nanotechnology for Solar Energy Collections and Conversion, Sustainable Nanomanufacturing—Creating the Industries of the Future, and Nanoelectronics for 2020 and Beyond. The report also emphasizes the importance of collaborative efforts, highlighting the role of the Nanotechnology Characterization Laboratory (NCL), a partnership between the National Cancer Institute, NIST, and FDA. The purpose of the NCL is to accelerate the transition of basic nanoscale particles and devices into clinical use by providing the necessary infrastructure and characterization services to nanomaterial developers.

The report also summarizes the stakeholder input it solicited in support of the development of the strategic plan. In particular, stakeholders helped to identify promising areas of nanotechnology research and proposed future nanotechnology signature initiatives; emphasized the importance of education and training to prepare U.S. scientists, engineers, technicians, and patent examiners;

---

[126] NSET Subcommittee, NSTC, EOP, *The National Nanotechnology Strategic Plan*, February 2011, http://www.nano.gov/nnistrategicplan211.pdf.

[127] NSET Subcommittee, NSTC, EOP, *The National Nanotechnology Strategic Plan*, December 2007, http://www.nano.gov/NNI_Strategic_Plan_2007.pdf.

identified successful models for technology transfer; and recommended strategies for increasing interagency collaboration.

Selected Nanotechnology Legislation in the 111ᵗʰ and 112ᵗʰ Congresses

S. 1662—Nanotechnology Regulatory Science Act of 2011

S. 1662, the Nanotechnology Regulatory Science Act of 2011, was introduced on October 6, 2011, and referred to the Senate Committee on Health, Education, Labor, and Pensions. The bill would amend the Federal Food, Drug, and Cosmetic Act (FFDCA) to require the Secretary of Health and Human Services to establish within the Food and Drug Administration a program for the scientific investigation of nanomaterials included or intended for inclusion in products regulated under the FFDCA to address the potential toxicology of such materials; the effects of such materials on biological systems; and the interaction of such materials with biological systems.

H.R. 2749—Nanotechnology Advancement and New Opportunities Act

H.R. 2749, the Nanotechnology Advancement and New Opportunities Act, was introduced on August 1, 2011, and referred to four House committees: the Committee on Science and Technology, the Committee on Energy and Commerce, the Committee on Ways and Means, and the Committee on Homeland Security. The purpose of the bill is to ensure the development and responsible stewardship of nanotechnology. The provisions of this bill are essentially identical to those in H.R. 820 (111ᵗʰ Congress). The bill would:

- establish a $100 million Nanomanufacturing Investment Partnership at the Department of Commerce to work with private investors to advance the commercialization of nanomanufacturing technologies and to increase the commercial application of federally supported research results;

- establish a 15% tax credit, taken over five years, for the purchase of up to $10 million of stock in qualified nanotechnology companies;

- establish a grant program within the DOC to support the establishment and development of nanotechnology incubators by non-profit entities and degree-granting institutions;

- require the NNCO Director to prepare a report to Congress on a nanotechnology research strategy for government and industry that will ensure the development and responsible stewardship of nanotechnology;

- provide a tax credit of 50% for nanotechnology education and training expenses for businesses and individuals;

- authorize an annual appropriation of $15 million for FY2012 through FY2015 for the NSF to conduct a grant program for the development of curriculum materials for interdisciplinary nanotechnology courses at institutions of higher education;

- direct the NSF to establish, through its Advanced Technological Education program, a program to encourage manufacturing companies to enter into partnerships with occupational training centers for the development of training to support nanomanufacturing; and

- direct the Secretary of Energy to submit a report to Congress containing a strategy for increasing interaction among scientists and engineers at DOE national laboratories and the informal science education community to prepare appropriate exhibits for school age children and the general public.

In addition, the bill would have amended the 21st Century Nanotechnology Research and Development Act of 2003 to:

- authorize $10 million for NSF to establish a center for the development of computer-aided design tools for nanotechnology applications;

- authorize an annual appropriation of $30 million for the DOE to conduct a grant program for nanotechnology research to address the need for "clean, cheap, renewable energy";

- authorize an annual appropriation of $30 million for the EPA for a grant program for nanotechnology research to address technologies for the remediation of pollution and other environmental protection technologies;

- authorize an annual appropriation of $30 million for the DHS to conduct a grant program for nanotechnology research to address the need for sensors and materials related to homeland security needs; and

- authorize an annual appropriation of $30 million for the DHHS to conduct a grant program for nanotechnology research to address health-related applications.

H.R. 2359—Safe Cosmetics Act of 2011

H.R. 2359, the Safe Cosmetics Act of 2011, was introduced on June 24, 2011, and referred to two House committees: the Committee on Energy and Commerce and the Committee on Education and the Workforce. The bill would amend the Federal Food, Drug, and Cosmetic Act to provide for the regulation of cosmetics by the Secretary of Health and Human Services. Among its provisions are several related to nanotechnology. Under the bill, the Secretary of HHS would be authorized to require that minerals and other particulate ingredients be labeled as "nano-scale" on a cosmetic ingredient label or list if not less than 1% of the ingredient particles in the cosmetic are 100 nanometers or smaller in not less than 1 dimension, and that other ingredients in a cosmetic be designated with scale-specific information on a cosmetic ingredient label or list if such ingredients possess scale-specific hazard properties. The bill would also require the Secretary of HHS to monitor developments in the scientific understanding of any adverse health effects related to the use of nanotechnology in the formulation of cosmetics and to consider scale specific hazard properties of ingredients when reviewing or evaluating the safety of cosmetics and ingredients. In addition, the bill would require manufacturers to submit to the Secretary a statement for each cosmetic that includes an ingredient list, including the particle size range of any nanoscale cosmetic ingredients.

S. 493—SBIR/STTR Reauthorization Act of 2011

S. 493, the SBIR/STTR Reauthorization Act of 2011, was introduced in the Senate on March 4, 2011, and referred to the Committee on Small Business and Entrepreneurship. The bill was reported with amendments but without a written report on March 9, 2011. On May 4, 2011, a cloture motion on the bill failed. As part of the larger purpose of reauthorizing the SBIR and STTR programs, S. 493 would require agencies with SBIR and STTR programs to give consideration to research topics identified in the NSTC/NSET's national nanotechnology strategic plan mandated by P.L. 108-153 and related documents, and to give special priority to applications for the support of projects related to nanotechnology and other specified fields of application. Section 501 (Research Topics and Program Diversification) of S. 493, which contained the nanotechnology related provisions, was incorporated in its entirety in S. 1867, the National Defense Authorization Act for Fiscal Year 2012. This bill was passed by the Senate, but the provision was not included in the final enacted version of the House bill (H.R. 1540, P.L. 112-81).

Title I, Subtitle A, H.R. 5116 (111th Congress)—National Nanotechnology Initiative Amendments Act of 2010

The provisions of Title I, Subtitle A of H.R. 5116 (111th Congress), the National Nanotechnology Initiative Amendments Act of 2010, were nearly identical to H.R. 554 (111th Congress) (see "H.R. 554 (111th Congress)—National Nanotechnology Initiative Amendments Act of 2009" below). H.R. 5116 changed the name of the act from the "National Nanotechnology Initiative Amendments Act of 2009," to "National Nanotechnology Initiative Amendments Act of 2010," and removed the term "interdisciplinary" from a provision establishing "green nanotechnology" research centers. The Senate removed this title before the bill was enacted.

H.R. 554 (111th Congress)—National Nanotechnology Initiative Amendments Act of 2009

H.R. 554 (111th Congress), the National Nanotechnology Initiative Amendments Act of 2009, was introduced on January 15, 2009, and passed by the House of Representatives on February 11, 2009. The bill was referred to the Senate Commerce, Science, and Transportation Committee on February 12, 2009. The purpose of the bill was to authorize activities for support of nanotechnology research and development and for other purposes. Among its provisions, the bill would have amended the 21st Century Nanotechnology Research and Development Act of 2003 to:

- require the NSTC triennial strategic plan to include near-term and long-term objectives, the anticipated timeframe for achieving near-term objectives, and metrics for assessing progress; cooperative and collaborative activities in R&D and technology transition supported by the states; and proposed research in areas of national priority;

- require the NSTC annual nanotechnology report supplementing the President's budget request to include a breakout of spending for the development and acquisition of research facilities and instrumentation for each program component area, and a breakout of spending on all activities related to ethical, legal, environmental, and societal implications;

- direct NNP agencies to support the activities of committees involved in the development of standards for nanotechnology and allow agencies to reimburse the travel costs of scientists and engineers who participate in activities of such committees;

- direct the agencies to fund the National Nanotechnology Coordination Office, and to do so in proportion to each agency's share of the previous year's NNP budget;

- require the NNCO to develop and maintain a publicly accessible database of projects funded under the Environmental, Health, and Safety, the Education and Societal Dimensions, and the Nanomanufacturing program component areas;

- require the NNCO to develop, maintain, and publicize information on nanotechnology facilities supported by the NNP, including at a minimum the terms and conditions for the use of each facility, a description of the capabilities of the instruments and equipment available for use at the facility, and a description of the technical support available to assist users of the facility;

- require the establishment of a National Nanotechnology Advisory Panel (NNAP) "as a distinct entity." Currently, under the provisions of presidential Executive Order 13349, the President's Council of Advisors on Science and Technology serves as the NNAP;[128]

- direct the NNCO to enter into an arrangement with the National Research Council to conduct a triennial review of the NNP, and authorizes funds for FY2010, FY2011, and FY2012; and

- define nanotechnology as "the science and technology that will enable one to understand, measure, manipulate, and manufacture at the nanoscale, aimed at creating materials, devices, and systems with fundamentally new properties or functions," and define nanoscale as "one or more dimensions of between approximately 1 and 100 nanometers."

In addition, the bill would have:

- required the designation of a White House Office of Science and Technology Policy associate director to serve as the "Coordinator for Societal Dimensions of Nanotechnology" and would charge the coordinator with convening and chairing a panel of federal agency representatives and others to develop, maintain, implement, and monitor an annual EHS research plan that includes, among other things, standards related to nanotechnology nomenclature; standards for methods and procedures for detecting, measuring, monitoring, sampling, and testing engineered nanoscale materials for environmental, health, and safety impacts; and standard reference materials for EHS testing;

- required the National Science Foundation to provide grants to establish Nanotechnology Education Partnerships to recruit and help prepare secondary school students to pursue postsecondary level courses of instruction in nanotechnology;

---

[128] Executive Order 13349, http://edocket.access.gpo.gov/cfr_2005/janqtr/3CFR13349 htm.

- directed the NSTC to establish an Education Working Group under the NSET Subcommittee to coordinate, prioritize, and plan NNP educational activities;

- directed certain NNP agencies to provide companies access to their supported facilities to assist in the development of prototypes of nanoscale products, devices, or processes for determining proof of concept;

- directed NNP agencies to encourage nanotechnology-related submissions to their Small Business Innovation Research (SBIR) and Small Business Technology Transfer (STTR) programs;

- directed NIST to encourage nanotechnology-related submissions to its Technology Innovation Program (TIP), and directs the TIP advisory Board to provide advice to NIST to accomplish this, and to provide an assessment of the adequacy of TIP resources allocated to nanotechnology related projects;

- directed the NSTC to actively pursue industry liaison groups for all industries;

- directed the NNP to coordinate and leverage federal investments with nanotechnology research, development, and technology transition initiatives supported by the States;

- directed the NNP to support nanotechnology R&D activities directed toward application areas that have the potential for significant contributions to national economic competitiveness and for other significant societal benefits, such as nano-electronics, energy efficiency, health care, and water remediation and purification;

- directed the NNP to support research on the development of instrumentation and tools required for the rapid characterization of nanoscale materials and for monitoring of nanoscale manufacturing processes, and to support approaches and techniques for scaling the synthesis of new nanoscale materials to achieve industrial-level production rates; and

- directed certain NNP-supported interdisciplinary research centers to support research on methods and approaches to environmentally benign nanoscale products and nanoscale manufacturing processes, as well as related technology transfer and education activities.

S. 1482 (111th Congress)—National Nanotechnology Amendments Act of 2009

S. 1482 (111th Congress), the National Nanotechnology Amendments Act of 2009, was introduced on July 21, 2009, and referred to the Senate Commerce, Science, and Transportation Committee. The purpose of the bill was to reauthorize the 21st Century Nanotechnology Research and Development Act and to expand the scope of the National Nanotechnology Program (NNP).

Among its provisions, the bill would have:

- required the NNP to solicit and draw upon the perspectives of the industrial community to promote the rapid commercial development of nanoscale-enabled devices, systems, and technologies and to coordinate research in determining the key physical and chemical characteristics of nanoparticles and nanomaterials that may pose environmental, health, and safety risks;

- required the NNCO and other appropriate agencies and councils to issue guidance to agencies that describes a strategy for transitioning research into commercial products and technologies and how the program will coordinate or conduct research on the environmental, health, and safety issues related to nanotechnology;

- required the NSTC triennial strategic plan to include near-term and long-term objectives, the anticipated timeframe for achieving near-term objectives, and metrics for assessing progress; cooperative and collaborative activities in R&D and technology transition supported by the states; how the NNP intends to encourage and support interdisciplinary research; and proposed research in areas of national priority;

- encouraged joint interagency solicitation of grant applications in high priority, multi-disciplinary research areas;

- required participating agencies to support the activities of the committees of standards setting bodies involved in the development of standards for nanotechnology;

- required each participating agency to provide funds to support the work of the NNCO. Authorizes appropriations to: (1) NIST for the development of nanotechnology standards; and (2) NSF, for use by the NNCO, to develop and maintain a public information database of NNP projects in EHS; education; public outreach; ethical, legal, and other societal issues; and of nanotechnology facilities accessible for use by individuals from academia and industry;

- made the National Nanotechnology Advisory Panel (NNAP) a distinct entity, and requires the NNAP to establish a subpanel to enable it to assess whether societal, ethical, legal, environmental, and workforce concerns are adequately addressed by the NNP;

- revised provisions for triennial external review of the NNP;

- required the designation of a "coordinator for societal dimensions of nanotechnology," within OSTP, to convene a panel to develop a research plan, and requires the coordinator to enter into an arrangement with the National Science Board to create a report that identifies the broad goals and needs of EHS researchers;

- directed the NSTC to establish an interagency Education Working Group to coordinate, prioritize, and plan formal and informal educational activities supported under the NNP, including activities to help participants understand the EHS implications of nanotechnology;

- provided for one or more grants to establish Nanotechnology Education Partnerships to recruit and help prepare secondary school students to pursue postsecondary level courses in nanotechnology;

- required agencies supporting nanotechnology research facilities to provide access to representatives from industry and other stakeholders for the transfer of research results or assist in developing proof-of-concept prototypes of nanoscale products, devices, or processes;

- directed NIST, in its Technology Innovation Program, and all agencies with Small Business Innovation Research (SBIR) and Small Business Technology Transfer (STTR) programs, to encourage the submission of nanotechnology related grant proposals;

- set, for the NNP, the objective of establishing industry liaison groups for all industry sectors that would benefit from nanotechnology applications;

- required coordination and leveraging of federal investments with nanotechnology research, development, and technology transition initiatives supported by state governments;

- required the NNP to support nanotechnology R&D in areas of national importance (e.g., economic competitiveness, energy production, water purification, agriculture, and health care; in environmental, health, and safety research on the risks of nanoparticles) and in ethical, legal, and societal issues related to nanotechnology;

- required the NNP to support a wide array of research in support of nanomanufacturing;

- required the director of the NNCO to review and report on nanomanufacturing research and research facilities;

- required an NNAP review of the nanomanufacturing program component area and the capabilities of nanotechnology research facilities supported by the NNP;

- set forth provisions regarding NNP nanoscale characterization and metrology research; and

- required deliberative public input in the decision making processes affecting policies for the research, development, and use of nanotechnology, and authorizes $2.0 million for the NNCO to carry out this responsibility.

S. 596 (111ᵗʰ Congress) — Nanotechnology Innovation and Prize Competition Act of 2009

S. 596 (111th Congress), the Nanotechnology Innovation and Prize Competition Act of 2009, was introduced on March 16, 2009, and referred to the Senate Commerce, Science, and Transportation Committee. The purpose of the bill was to establish an award program to honor achievements in nanotechnology. Under the bill, the Department of Commerce's National Institute of Standards and Technology is directed to award prizes to individuals and companies for achievement in one or more of the following areas: improvement of the environment, consistent with EPA's Twelve Principles of Green Chemistry; development of alternative energy that has the potential to lessen the dependence of the United States on fossil fuels; and/or improvement of human health, consistent with regulations promulgated by the FDA. The bill would have authorized financial prizes for being the first to achieve a specific criteria, as well as recognition prizes, made as part of the previously established National Technology and Innovation Medal award program. The bill would have authorized $2 million annually for the financial prizes as well as $750,000 annually for administration of the prize competitions.

H.R. 820 (111ᵗʰ Congress)—Nanotechnology Advancement and New Opportunities Act

H.R. 820 (111ᵗʰ Congress), the Nanotechnology Advancement and New Opportunities Act, was introduced on February 3, 2009, and referred to four House committees: the Committee on Science and Technology, the Committee on Energy and Commerce, the Committee on Ways and Means, and the Committee on Homeland Security. The purpose of the bill was to ensure the development and responsible stewardship of nanotechnology. The provisions of this bill are nearly identical to those of H.R. 2749, introduced in the 112ᵗʰ Congress (described above).

H.R. 2647 (111ᵗʰ Congress)—National Defense Authorization Act for Fiscal Year 2010

Section 242 of the National Defense Authorization Act for Fiscal Year 2010 (H.R. 2647, P.L. 111-84) amends the Department of Defense's nanotechnology reporting responsibilities to align with those required of other agencies under the 21ˢᵗ Century Nanotechnology Research and Development Act (P.L. 108-153). H.R. 2647 was signed into law on October 28, 2009.

S. 3117 (111ᵗʰ Congress)—Promote Nanotechnology in Schools Act

S. 3117 (111ᵗʰ Congress), the Promote Nanotechnology in Schools Act, was introduced on March 15, 2010, and referred to the Senate Committee on Health, Education, Labor, and Pensions. The purpose of the bill was to strengthen the capacity of eligible institutions (i.e., secondary schools, community colleges, two-year and four-year institutions of higher education, and informal learning science centers) to provide instruction in nanotechnology. The bill would have authorized a program at the National Science Foundation for this purpose that would offer eligible institutions grants of up to $400,000 (subject to a 25% match from non-federal sources) to assist in the purchase and maintenance of nanotechnology equipment and software, to develop and provide educational services, and to support teacher education and certification. The bill would have authorized $15 million for FY2010 and "such sums as may be necessary" for FY2011 through FY 2013.

H.R. 4502 (111ᵗʰ Congress)—Nanotechnology Education Act

H.R. 4502 (111ᵗʰ Congress), the Nanotechnology Education Act, was introduced on February 19, 2010, and referred to the House Committee on Science and Technology's Subcommittee on Research and Science Education. The purpose of the bill was to strengthen the capacity of eligible institutions (i.e., secondary schools, community colleges, four-year institutions of higher education, and informal learning science centers) to provide instruction in nanotechnology. The bill would have authorized a program for this purpose at the National Science Foundation that would offer eligible institutions grants of up to $400,000 (subject to a 25% match from non-federal sources) to assist in the purchase and maintenance of nanotechnology equipment and software, to develop and provide educational services, and to support teacher education and certification. The bill would have authorized $40 million for FY2011 and "such sums as may be necessary" for FY2012 through FY 2014.

S. 2942 (111ᵗʰ Congress) — Nanotechnology Safety Act of 2010

S. 2942 (111ᵗʰ Congress), the Nanotechnology Safety Act of 2010, was introduced on January 21, 2010, and referred to the Senate Committee on Health, Education, Labor, and Pensions. The bill would have required the Secretary of Health and Human Services to establish within 180 days a program for the scientific investigation of nanoscale materials included or intended for inclusion in FDA-regulated products, to address the potential toxicology of such materials, the effects of such materials on biological systems, and interaction of such materials with biological systems. The bill would have authorized $25 million per year for fiscal years 2011 to 2015.

H.R. 5786 (111ᵗʰ Congress) — Safe Cosmetics Act of 2010

H.R. 5786 (111ᵗʰ Congress), the Safe Cosmetics Act of 2010, was introduced on July 20, 2010, and referred to the House Committee on Energy and Commerce and the House Committee on Education and Labor. The bill would have required the Secretary of Health and Human Services to monitor developments in the scientific understanding of any adverse health effects related to the use of nanotechnology in the formulation of cosmetics and to consider scale specific hazard properties of ingredients when conducting or reviewing safety substantiation of cosmetic ingredients. In addition, the bill would have required manufacturers to submit to the Secretary a statement for each cosmetic that includes an ingredient list, including the particle size of any nanoscale cosmetic ingredients. The bill would also have given the Secretary authority to require labeling of cosmetics disclosing the use of nanoscale ingredients.

Concluding Observations

Many expect nanotechnology to bring significant economic and societal returns. The United States was the first government to launch a national-level nanotechnology program and has invested more than any other nation. As a result of this focus and these sustained investments, many experts believe that the United States enjoys a technological leadership position in nanotechnology. Other nations are investing heavily and some industrialized and emerging economies have formidable capabilities in nanotechnology. Assessments of the National Nanotechnology Initiative have concluded that the effort is well-managed and has been successful in achieving its objectives so far. However, these assessments have recognized that the NNI faces a variety of challenges in ensuring that the full promise of nanotechnology is realized and that the United States remains the global leader in nanoscale science, engineering, and technology.

Congress may choose to address some or many of the issues addressed in the body of this report in the course of deliberation on the reauthorization of the 21ˢᵗ Century Nanotechnology R&D Act of 2003 or, alternatively, in separate legislation.

The 21ˢᵗ Century Nanotechnology R&D Act's funding authorizations extended through FY2008. The 109ᵗʰ Congress, 110ᵗʰ Congress, and 111ᵗʰ Congress, considered legislation to reauthorize the program. If the 112ᵗʰ Congress opts to consider reauthorization of the act, some of the issues it may wish to consider include budget authorization levels for the covered agencies; R&D funding levels, priorities, and balance across the program component areas; administration and management of the NNI; translation of research results and early-stage technology into commercially viable applications; environmental, health, and safety issues; ethical, legal, and societal implications; education and training for the nanotechnology workforce; metrology, standards, and nomenclature; public understanding; and international dimensions. Consideration may also be given to the establishment of an independent review panel and to coordination of the timing for the NNAP assessment, the NRC assessment, and the NSET Subcommittee's strategic plan for the NNI.

Appendix A. Selected Reports on the National Nanotechnology Initiative

Reports of the Nanoscale Science, Engineering, and Technology Subcommittee of the National Science and Technology Council

The National Nanotechnology Initiative: Research and Development Leading to a Revolution in Technology and Industry, Supplement to the President's FY2013 Budget, February 2012. http://nano.gov/sites/default/files/pub_resource/nni_2013_budget_supplement.pdf

Environmental, Health, and Safety Research Strategy, October 2011. http://nano.gov/sites/default/files/pub_resource/nni_2011_ehs_research_strategy.pdf

Policy Principles for the U.S. Decision-Making Concerning Regulation and Oversight of Applications of Nanotechnology and Nanomaterials, June 2011. http://www.whitehouse.gov/sites/default/files/omb/inforeg/for-agencies/nanotechnology-regulation-and-oversight-principles.pdf

The National Nanotechnology Initiative: Research and Development Leading to a Revolution in Technology and Industry, Supplement to the President's FY2012 Budget, March 2011. http://nano.gov/sites/default/files/pub_resource/nni_2012_budget_supplement.pdf

Regional, State, and Local Initiatives in Nanotechnology: Report of the National Nanotechnology Initiative Workshop, February 2011. http://nano.gov/sites/default/files/pub_resource/nni_2012_budget_supplement.pdf

National Nanotechnology Initiative Strategic Plan, February 2011. http://nano.gov/sites/default/files/pub_resource/2011_strategic_plan.pdf

National Nanotechnology Initiative Signature Initiative: Nanotechnology for Solar Energy Collection and Conversion, July 2010. http://nano.gov/sites/default/files/pub_resource/nnisiginitsolarenergyfinaljuly2010.pdf

National Nanotechnology Initiative Signature Initiative: Sustainable Nanomanufacturing – Creating the Industries of the Future, July 2010. http://nano.gov/sites/default/files/pub_resource/nni_siginit_sustainable_mfr_revised_nov_2011.pdf

National Nanotechnology Initiative Signature Initiative: Nanoelectronics for 2020 and Beyond, July 2010. http://nano.gov/sites/default/files/pub_resource/nni_siginit_nanoelectronics_jul_2010.pdf

nanoEHS Series: Risk Management Methods & Ethical, Legal, and Societal Implications of Nanotechnology: Report of the National Nanotechnology Initiative Workshop, March 2010.

Communicating Risk in the 21st Century: The Case of Nanotechnology, February 2010. http://nano.gov/sites/default/files/pub_resource/berube_risk_white_paper_feb_2010.pdf

The National Nanotechnology Initiative: Research and Development Leading to a Revolution in Technology and Industry, Supplement to the President's FY2011 Budget, February 2010. nano.gov/sites/default/files/pub_resource/nni_2011_budget_supplement.pdf

nanoEHS Series: Nanomaterials and Human Health & Instrumentation, Metrology, and Analytical Methods: Report of the National Nanotechnology Initiative Workshop, November 2009. http://nano.gov/sites/default/files/pub_resource/nanoandhumanhealthandinstrumentation.pdf

The National Nanotechnology Initiative: Research and Development Leading to a Revolution in Technology and Industry, Supplement to the President's FY2010 Budget, May 2009. http://nano.gov/sites/default/files/pub_resource/nni_2010_budget_supplement.pdf

Nanotechnology-Enabled Sensing, Report of the National Nanotechnology Initiative Workshop, May 2009. http://nano.gov/sites/default/files/NNI-Nanosensors-stdres.pdf

nanoEHS Series: Human and Environmental Exposure Assessment Workshop Materials, February 2009. http://nano.gov/node/122

The National Nanotechnology Initiative: Research and Development Leading to a Revolution in Technology and Industry, Supplement to the President's FY2009 Budget, September 2008. http://nano.gov/sites/default/files/pub_resource/nni_09budget.pdf

Strategy for Nanotechnology-Related Environmental, Health, and Safety Research, February 2008. http://nano.gov/sites/default/files/pub_resource/nni_ehs_research_strategy.pdf

The National Nanotechnology Initiative Strategic Plan, December 2007. http://nano.gov/sites/default/files/pub_resource/nni_strategic_plan_2007.pdf

Prioritization of Environmental, Health, and Safety Research Needs for Engineered Nanoscale Materials, August 2007. http://nano.gov/sites/default/files/pub_resource/prioritization_ehs_research_needs_engineered_nanoscale_materials.pdf

The National Nanotechnology Initiative: Research and Development Leading to a Revolution in Technology and Industry, Supplement to the President's FY2008 Budget, July 2007. http://nano.gov/sites/default/files/pub_resource/nni_08budget.pdf

Manufacturing at the Nanoscale: Report of the National Nanotechnology Initiative Workshop, January 2007. http://nano.gov/sites/default/files/pub_resource/manufacturing_at_the_nanoscale.pdf

The National Nanotechnology Initiative: Environmental, Health, and Safety Research Needs for Engineered Nanoscale Materials, September 2006. http://nano.gov/sites/default/files/pub_resource/nni_ehs_research_needs.pdf

The National Nanotechnology Initiative: Research and Development Leading to a Revolution in Technology and Industry, Supplement to the President's FY2007 Budget, July 2006. http://nano.gov/sites/default/files/pub_resource/nni_07budget.pdf

The National Nanotechnology Initiative: Research and Development Leading to a Revolution in Technology and Industry, Supplement to the President's FY2006 Budget, March 2005. http://nano.gov/sites/default/files/pub_resource/nni_06budget.pdf

The National Nanotechnology Initiative Strategic Plan, December 2004.

Research Directions II: Long-Term Research and Development Opportunities in Nanotechnology, Report of the National Nanotechnology Initiative Workshop, September 2004. http://nano.gov/sites/default/files/pub_resource/research_directionsii.pdf

Nanotechnology in Space Exploration, August 2004. http://nano.gov/sites/default/files/pub_resource/space_exploration_rpt_0.pdf

Nanoscience Research for Energy Needs: Report of the National Nanotechnology Initiative Grand Challenge Workshop, March 2004. http://nano.gov/sites/default/files/nni_energy_rpt.pdf

Nanoelectronics, Nanophotonics, & Nanomagnetics: Report of the National Nanotechnology Initiative Workshop, February 2004. http://nano.gov/sites/default/files/pub_resource/nni_electronic_photonics_m.pdf

Instrumentation and Metrology for Nanotechnology: Report of the National Nanotechnology Initiative Workshop, January 2004. http://nano.gov/sites/default/files/pub_resource/nni_instrumentation_metrology_rpt.pdf

Nanotechnology: Societal Implications-Maximizing Benefits for Humanity: Report of the National Nanotechnology Initiative Workshop, December 2003. http://nano.gov/sites/default/files/nni_societal_implications.pdf

Nanobiotechnology: Report of the National Nanotechnology Initiative Workshop, October 2003. http://nano.gov/sites/default/files/pub_resource/nni_nanobiotechnology_rpt.pdf

Regional, State, and Local Initiatives in Nanotechnology, September-October 2003. [No URL available.]

National Nanotechnology Initiative: Research and Development Supporting the Next Industrial Revolution, Supplement to the President's FY2004 Budget. August 2003. http://nano.gov/sites/default/files/pub_resource/nni04_budget_supplement.pdf

Materials by Design: Report of the National Nanotechnology Initiative Workshop, June 2003. http://nano.gov/sites/default/files/pub_resource/nni_materials_by_design.pdf

Nanotechnology and the Environment: Report of the National Nanotechnology Initiative Workshop, May 2003. http://nano.gov/sites/default/files/pub_resource/nanotechnology_and_the_environment_app_imp.pdf

National Nanotechnology Initiative: The Initiative and Its Implementation Plan, Detailed Technical Report Associated with the Supplemental Report to the President's FY2003 Budget, June 2002.

Societal Implications of Nanoscience and Nanotechnology, March 2001. http://www.wtec.org/loyola/nano/NSET.Societal.Implications/report-grayscale.pdf

National Nanotechnology Initiative: The Initiative and Its Implementation Plan, Detailed Technical Report Associated with the Supplemental Report to the President's FY2001 Budget, July 2000. http://nano.gov/sites/default/files/pub_resource/nni_implementation_plan_2000.pdf

Report of the Interagency Working Group on Nanoscience, Technology, and Engineering (NSET Subcommittee Predecessor)

Nanotechnology Research Directions, IWGN Workshop Report, September 1999. http://nano.gov/sites/default/files/pub_resource/research_directions_1999.pdf

Nanotechnology: Shaping the World Atom by Atom, 1999. http://www.wtec.org/loyola/nano/IWGN.Public.Brochure/IWGN.Nanotechnology.Brochure.pdf

Agency Reports

Defense Nanotechnology Research and Development Program, Office of the Director of Defense Research and Engineering, Department of Defense, December 2009. http://nano.gov/sites/default/files/pub_resource/dod-report_to_congress_final_1mar10.pdf

Current Intelligence Bulletin 60: Interim Guidance for Medical Screening and Hazard Surveillance for Workers Potentially Exposed to Engineered Nanoparticles, National Institute for Occupational Safety and Health, Centers for Disease Control and Prevention, Department of Health and Human Services. February 2009. http://www.cdc.gov/niosh/docs/2009-116/pdfs/2009-116.pdf

Progress Toward Safe Nanotechnology in the Workplace: A Report from the NIOSH Nanotechnology Research Center, National Institute for Occupational Safety and Health, Centers for Disease Control and Prevention, Department of Health and Human Services, June 2007. http://www.cdc.gov/niosh/docs/2010-104/pdfs/2010-104.pdf

Approaches to Safe Nanotechnology in the Workplace: Managing the Health and Safety Concerns Associated with Engineering Nanomaterials, National Institute for Occupational Safety and Health, Centers for Disease Control and Prevention, Department of Health and Human Services, March 2009.

External Reviews

Report to the President and Congress on the Fourth Assessment of the National Nanotechnology Initiative, PCAST (acting as the NNAP), April 2012. http://www.whitehouse.gov/sites/default/files/microsites/ostp/PCAST_2012_Nanotechnology_FINAL.pdf

Review of Federal Strategy for Nanotechnology-Related Environmental, Health, and Safety Research, Committee for Review of the Federal Strategy to Address Environmental, Health, and Safety Research Needs for Engineered Nanoscale Materials, Committee on Toxicology, NRC, 2009. http://www.nap.edu/catalog.php?record_id=12559#toc

Report to the President and Congress on the Third Assessment of the National Nanotechnology Initiative, PCAST (acting as the NNAP), March 2010. http://www.whitehouse.gov/sites/default/files/microsites/ostp/pcast-nni-report.pdf

The National Nanotechnology Initiative: Second Assessment and Recommendations of the National Nanotechnology Advisory Panel, PCAST (acting as the NNAP), April 2008.

http://www.whitehouse.gov/sites/default/files/microsites/ostp/PCAST-NNAP-NNI-Assessment-2008.pdf

A Matter of Size: Triennial Review of the National Nanotechnology Initiative, NRC, 2006. http://www.nap.edu/catalog.php?record_id=11752

The National Nanotechnology Initiative at Five Years: Assessment and Recommendations of the National Nanotechnology Advisory Panel, PCAST (acting as the NNAP), May 2005. http://www.whitehouse.gov/sites/default/files/microsites/ostp/pcast-nni-five-years.pdf

Small Wonders, Endless Frontiers: A Review of the National Nanotechnology Initiative, NRC, June 2002. http://www.nap.edu/openbook.php?isbn=0309084547

Appendix B. List of NNI and Nanotechnology-Related Acronyms

ASTRA	Alliance for Science and Technology Research in America
CNST	Center for Nanoscale Science and Technology
CS	Committee on Science
CT	Committee on Technology
CSREES	Cooperative State Research, Education, and Extension Service
DHHS	Department of Health and Human Services
DHS	Department of Homeland Security
DOC	Department of Commerce
DOD	Department of Defense
DOE	Department of Energy
DOJ	Department of Justice
DOT	Department of Transportation
EHS	Environmental, health, and safety
ELSI	Ethical, legal, and societal implications
EPA	Environmental Protection Agency
EOP	Executive Office of the President
EPSCoR	Experimental Program to Stimulate Competitive Research
FFDCA	Federal Food, Drug, and Cosmetic Act
FHWA	Federal Highway Administration
GIN	Global Issues in Nanotechnology working group
ISO	International Standards Organization
IWGN	Interagency Working Group on Nanotechnology
NASA	National Aeronautics and Space Administration
NCI	National Cancer Institute
NEHI	National Environmental and Health Implications working group
NGO	Non-governmental organization
NIH	National Institutes of Health
NILI	National Innovation and Liaison with Industry working group
NIOSH	National Institute of Occupational Safety and Health
NIST	National Institute of Standards and Technology
NNAP	National Nanotechnology Advisory Panel
NNCO	National Nanotechnology Coordination Office
NNI	National Nanotechnology Initiative
NNIN	National Nanotechnology Infrastructure Network
NNN	National Nanomanufacturing Network

NNP	National Nanotechnology Program
NPEC	Nanotechnology Public Engagement and Communications working group
NRC	National Research Council
NSET	Nanoscale Science, Engineering and Technology subcommittee
NSF	National Science Foundation
NSEC	Nanoscale Science and Engineering Center
NSI	Nanotechnology Signature Initiative
NSRC	Nanoscale Science Research Centers
NSTC	National Science and Technology Council
OECD	Organization for Economic Cooperation and Development
OMB	Office of Management and Budget
OSTP	Office of Science and Technology Policy
PCA	Program Component Areas
PCAST	President's Council of Advisors on Science and Technology
R&D	Research and development
SBIR	Small Business Innovation Research
STTR	Small Business Technology Transfer Research
TSA	Transportation Safety Administration
USDA	U.S. Department of Agriculture
USPTO	U.S. Patent and Trademark Office

Author Contact Information

John F. Sargent Jr.
Specialist in Science and Technology Policy
jsargent@crs.loc.gov, 7-9147